An Introduction
to Genetic Algorithms
for Scientists and Engineers

An Introduction
to Genetic Algorithms
for Scientists and Engineers

David A Coley

University of Exeter

 **World Scientific**
Singapore • New Jersey • London • Hong Kong

Published by

World Scientific Publishing Co. Pte. Ltd.

5 Toh Tuck Link, Singapore 596224

USA office: 27 Warren Street, Suite 401-402, Hackensack, NJ 07601

UK office: 57 Shelton Street, Covent Garden, London WC2H 9HE

British Library Cataloguing-in-Publication Data
A catalogue record for this book is available from the British Library.

First published 1999
Reprinted 2001, 2003, 2005

ISBN 981-02-3602-6

Printed in Singapore by Utopia Press Pte Ltd

In the beginning was the Word
And by the mutations came the Gene

M.A. Arbid

Word
Wore
Gore
Gone
Gene

To my parents

PREFACE

Genetic algorithms (GAs) are general search and optimisation algorithms inspired by processes normally associated with the natural world. The approach is gaining a growing following in the physical, life, computer and social sciences and in engineering. Typically those interested in GAs can be placed into one or more of three rather loose categories:

1. those using such algorithms to help understand some of the processes and dynamics of natural evolution;
2. computer scientists primarily interested in understanding and improving the techniques involved in such approaches, or constructing advanced adaptive systems; and
3. those with other interests, who are simply using GAs as a way to help solve a range of difficult modelling problems.

This book is designed first and foremost with this last group in mind, and hence the approach taken is largely practical. Algorithms are presented in full, and working code (in BASIC, FORTRAN, PASCAL and C) is included on a floppy disk to help you to get up and running as quickly as possible. Those wishing to gain a greater insight into the current computer science of GAs, or into how such algorithms are being used to help answer questions about natural evolutionary systems, should investigate one or more of the texts listed in Appendix A.

Although I place myself in the third category, I do find there is something fascinating about such evolutionary approaches in their own right, something almost seductive, something fun. Why this should be I do not know, but there is something incredible about the power of the approach that draws one in and creates a desire to know that little bit more and a wish to try it on ever harder problems.

All I can say is this: if you have never tried evolutionary inspired methods before, you should suspend your disbelief, give it a go and enjoy the ride.

This book has been designed to be useful to most practising scientists and engineers (not necessarily academics), whatever their field and however rusty their mathematics and programming might be. The text has been set at an introductory, undergraduate level and the first five chapters could be used as part of a taught course on search and optimisation. Because most of the operations and processes used by GAs are found in many other computing situations, for example: loops; file access; the sorting of lists; transformations; random numbers; the systematic adjustment of internal parameters; the use of multiple runs to produce statistically significant results; and the role of stochastic errors, it would, with skill, be possible to use the book as part of a general scientific or engineering computing course. The writing of a GA itself possibly makes an ideal undergraduate exercise, and its use to solve a real engineering or scientific problem a good piece of project work. Because the algorithm naturally separates into a series of smaller algorithms, a GA could also form the basis of a simple piece of group programming work.

Student exercises are included at the end of several of the chapters. Many of these are computer-based and designed to encourage an exploration of the method.

Please email any corrections, comments or questions to the address below. Any changes to the text or software will be posted at http://www.ex.ac.uk/cee/ga/ .

David A. Coley
Physics Department
University of Exeter
September 1998

D.A.Coley@exeter.ac.uk

All trademarks are acknowledged as the property of their respective owners

The Internet and World Wide Web addresses were checked as close to publication as possible, however the locations and contents of such sites are subject to changes outwith the control of the author. The author also bears no responsibility for the contents of those web-pages listed, these addresses are given for the convenience and interest of readers only.

ACKNOWLEDGEMENTS

As with most books, this one owes its existence to many people. First and foremost to my wife Helen, who translated my scribbled notes and typed up the result; but more importantly, for organising the helicopters and ambulances, and looking after my body after I lost one of many arguments with gravity.

Many colleagues have helped directly with the text, either with the research, providing material for the text, or reading the drafts; including: Thorsten Wanschura, Stefan Migowsky, David Carroll, Godfrey Walters, Dragan Savic, Dominic Mikulin, Andrew Mitchell and Richard Lim of World Scientific.

A final word of thanks is owed to Mr Lawson, my school science teacher, whose enthusiasm and teaching first opened my eyes to the wonders of the physical and natural world and the possibilities of science.

NOTATION

$\oplus$	Concatenation, e.g. $00 \oplus 11 = 0011$.
C	A chromosome, string or structure.
C^*	The form of the chromosome, string or structure at the global optimum.
c_m	Fitness scaling constant.
d	Defining length of a schema.
f	Fitness.
f^{share}	The shared fitness.
f^*	Fitness at the global optimum.
f_{ave}	The average population fitness.
f_{max}	The fitness of the elite member.
f_{min}	The minimum fitness required for a solution to be defined as acceptable, e.g. the lower boundary of the fitness filter when hunting for a series of approximate solutions.
f_{off}	The off-line system performance.
f_{on}	The on-line system performance.
f_{sum}	Sum of the fitness of all population members in the current generation.
f_σ	The standard deviation of the population fitness in the current generation.
g	Generation.
G	Maximum number of generations.
i	Typically used to indicate an individual within the population.
l	Length of the binary string used to represent a single unknown.
L	Total string (or chromosome) length.
M	Number of unknown parameters in the problem.
$MAX[t(x)]$	Find the maximum of function t.
N	Population size.
o	Order of a schema.
P_c	Crossover probability.
P_m	Mutation probability.
R	Random decimal number.
r	Unknown parameter (typically real-valued), the optimum value of which the GA is being used to discover.
$\acute{r}$	Fitness-proportional (roulette wheel) selection.
R^+	Random decimal number in the range 0 to +1.
$R^\pm$	Random decimal number in the range -1 to +1.

R_c	Random number used to decide if an individual will undergo crossover.
R_L	Random number indicating the crossover position.
r_{max}	Maximum possible value of the unknown r.
r_{min}	Minimum possible value of the unknown r.
R_s	Random number used to decide if an individual will be selected to go forward to the next generation.
S	Schema.
s_i	Sharing function for individual i.
x	Variable; c.f. x, which indicates multiplication.
z	Unknown parameter expressed as a base ten integer.
z_{max}	Maximum value of an unknown expressed as a base ten integer.
z_{min}	Minimum value of an unknown expressed as a base ten integer.
$\acute{\varepsilon}$	Elitism; equals 1 if elitism is being applied, 0 if not.
η_{max}	Genotypic similarity between the elite member and the rest of the population.
ξ_{ij}	Sharing value between individuals i and j.
ρ	The number of optima in a multimodal function.
τ	Actual number of trials in the next generation.
τ^{ave}	Number of trials in the next generation an individual of average fitness might receive.
τ^{best}	Number of trials in the next generation the elite member will receive.
τ^{exp}	Expected number of trials in the next generation.
τ_{best}^{exp}	Expected number of trials in the next generation the elite member might receive.
Γ	The take-over time, i.e., the number of generations taken by the elite member to dominate the population.
φ	Selection mechanism.
Ω	Problem objective function (the maximum or minimum value of which is being sought), fitness will typically be a very simple function of Ω.
Φ	Number of instances of a particular schema within a population.
$Я$	The number of multiple runs carried out to reduce the effects of stochastic errors.
101	Example of a binary number; c.f. 101, a decimal number.

CONTENTS

CHAPTER 1

INTRODUCTION

Genetic algorithms (GAs) are numerical optimisation algorithms inspired by both natural selection and natural genetics. The method is a general one, capable of being applied to an extremely wide range of problems. Unlike some approaches, their promise has rarely been over-sold and they are being used to help solve practical problems on a daily basis. The algorithms are simple to understand and the required computer code easy to write. Although there is a growing number of disciples of GAs, the technique has never attracted the attention that, for example, artificial neural networks have. Why this should be is difficult to say. It is certainly not because of any inherent limits or for lack of a powerful metaphor. What could be more inspiring than generalising the ideas of Darwin and others to help solve other real-world problems? The concept that evolution, starting from not much more than a chemical "mess", generated the (unfortunately vanishing) bio-diversity we see around us today is a powerful, if not awe-inspiring, paradigm for solving any complex problem.

In many ways the thought of extending the concept of natural selection and natural genetics to other problems is such an obvious one that one might be left wondering why it was not tried earlier. In fact it was. From the very beginning, computer scientists have had visions of systems that mimicked one or more of the attributes of life. The idea of using a population of solutions to solve practical engineering optimisation problems was considered several times during the 1950's and 1960's. However, GAs were in essence invented by one man—John Holland—in the 1960's. His reasons for developing such algorithms went far beyond the type of problem solving with which this text is concerned. His 1975 book, *Adaptation in Natural and Artificial Systems* [HO75] (recently re-issued with additions) is particularly worth reading for its visionary approach. More recently others, for example De Jong, in a paper entitled *Genetic Algorithms are NOT Function Optimizers* [DE93], have been keen to remind us that GAs are potentially far more than just a robust method for estimating a series of unknown parameters within a model of a physical system. However in the context of this text, it is this robustness across many different practical optimisation problems that concerns us most.

So what is a GA? A typical algorithm might consist of the following:

1. a number, or population, of guesses of the solution to the problem;

2. a way of calculating how good or bad the individual solutions within the population are;

3. a method for mixing fragments of the better solutions to form new, on average even better solutions; and

4. a mutation operator to avoid permanent loss of diversity within the solutions.

With typically so few components, it is possible to start to get the idea of just how simple it is to produce a GA to solve a specific problem. There are no complex mathematics, or torturous, impenetrable algorithms. However, the downside of this is that there are few hard and fast rules to what exactly a GA is.

Before proceeding further and discussing the various ways in which GAs have been constructed, a sample of the range of the problems to which they have been successfully applied will be presented, and an indication given of what is meant by the phrase "search and optimisation".

1.1 SOME APPLICATIONS OF GENETIC ALGORITHMS

Why attempt to use a GA rather than a more traditional method? One answer to this is simply that GAs have proved themselves capable of solving many large complex problems where other methods have experienced difficulties. Examples are large-scale combinatorial optimisation problems (such as gas pipe layouts) and real-valued parameter estimations (such as image registrations) within complex search spaces riddled with many local optima. It is this ability to tackle search spaces with many local optima that is one of the main reasons for an increasing number of scientists and engineers using such algorithms.

Amongst the many practical problems and areas to which GAs have been successfully applied are:

- image processing [CH97,KA97];
- prediction of three dimensional protein structures [SC92];
- VLSI (very large scale integration) electronic chip layouts [COH91,ES94];
- laser technology [CA96a,CA96b];
- medicine [YA98];
- spacecraft trajectories [RA96];
- analysis of time series [MA96,ME92,ME92a,PA90];
- solid-state physics [SU94,WA96];
- aeronautics [BR89,YA95];
- liquid crystals [MIK97];
- robotics [ZA97, p161-202];
- water networks [HA97,SA97];
- evolving cellular automaton rules [PA88,MI93,MI94a];
- the architectural aspects of building design [MIG95,FU93];
- the automatic evolution of computer software [KO91,KO92,KO94];
- aesthetics [CO97a];
- jobshop scheduling [KO95,NA91,YA95];
- facial recognition [CA91];
- training and designing artificial intelligence systems such as artificial neural networks [ZA97, p99-117,WH92, KI90,KI94,CH90]; and
- control [NO91,CH96,CO97].

1.2 SEARCH SPACES

In a numerical search or optimisation problem, a list, quite possibly of infinite length, of possible solutions is being searched in order to locate the solution that best describes the problem at hand. An example might be trying to find the best values for a set of adjustable parameters (or variables) that, when included in a mathematical model, maximise the lift generated by an aeroplane's wing. If there were only two of these adjustable parameters, a and b, one could try a large number of combinations, calculate the lift generated by each design and produce a surface plot with a, b and *lift* plotted on the x-, y- and z-axis respectively (Figure 1.0). Such a plot is a representation of the problem's *search space*. For more complex problems, with more than two unknowns, the situation becomes harder to visualise. However, the concept of a search space is still valid as long as some measure of distance between solutions can be defined and each solution can be assigned a measure of success, or *fitness*, within the problem. Better performing, or fitter, solutions will then occupy the peaks within the search space (or *fitness landscape* [WR31]) and poorer solutions the valleys.

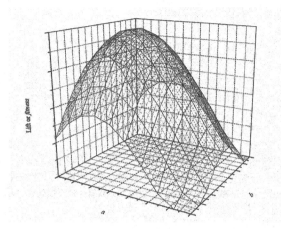

Figure 1.0. A simple search space or "fitness landscape". The lift generated by the wing is a function of the two adjustable parameters *a* and *b*. Those combinations which generate more lift are assigned a higher fitness. Typically, the desire is to find the combination of the adjustable parameters that gives the highest fitness.

Such spaces or landscapes can be of surprisingly complex topography. Even for simple problems, there can be numerous peaks of varying heights, separated from each other by valleys on all scales. The highest peak is usually referred to as the *global maximum* or *global optimum*, the lesser peaks as *local maxima* or *local optima*. For most search problems, the goal is the accurate identification of the global optimum, but this need not be so. In some situations, for example real-time control, the identification of **any** point above a certain value of fitness might be acceptable. For other problems, for example, in architectural design, the identification of a large number of highly fit, yet distant and therefore distinct, solutions (designs) might be required.

To see why many traditional algorithms can encounter difficulties, when searching such spaces for the global optimum, requires an understanding of how the features within spaces are formed. Consider the experimental data shown in Figure 1.1, where measurements of a dependent variable y have been made at various points j of the independent variable x. Clearly there is some evidence that x and y might be related through:

$$y_j = mx_j + c \ .$$
(1.1)

But what values should be given to m and c? If there is reason to believe that $y = 0$ when $x = 0$ (i.e. the line passes through the origin) then $c = 0$ and m is the only adjustable parameter (or *unknown*).

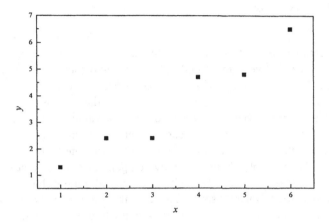

Figure 1.1. Some simple experimental data possibly related by $y = mx + c$.

One way of then finding m is simply to use a ruler and estimate the best line through the points by eye. The value of m is then given by the slope of the line. However there are more accurate approaches. A common numerical way of finding the best estimate of m is by use of a least-squares estimation. In this technique the error between that y predicted using (1.1) and that measured during the experiment, $\bar{y}$, is characterised by the objective function, Ω, (in this case the least squares cost function) given by,

$$\Omega = \sum_{j=1}^{n} \left(\bar{y}_j - y_j \right)^2$$
(1.2)

where n is the number of data points. Expanding (1.2) gives:

$$\Omega = \sum_{j=1}^{n} \left(\bar{y}_j - \left(mx_j + c \right) \right)^2 .$$

As $c = 0$,

$$\Omega = \sum_{j=1}^{n} \left(\bar{y}_j - mx_j \right)^2 . \tag{1.3}$$

In essence the method simply calculates the sum of the squares of the vertical distances between measured values of y and those predicted by (1.1) (see Figure 1.2). Ω will be at a minimum when these distances sum to a minimum. The value of m which gives this value is then the best estimate of m. This still leaves the problem of finding the lowest value of Ω. One way to do this (and a quite reasonable approach given such an easy problem with relatively few data points) is to use a computer to calculate Ω over a fine grid of values of m. Then simply choose the m which generates the lowest value of Ω. This approach was used together with the data of Figure 1.1 to produce a visualisation of the problem's search space—Figure 1.3. Clearly, the best value of m is given by $m = m^* \approx 1.1$, the asterisk indicating the optimal value of the parameter.

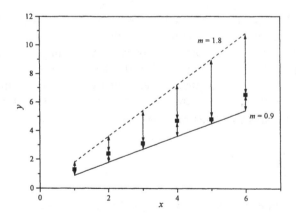

Figure 1.2. Calculating Ω for two values of m. Clearly $m = 0.9$ is the better choice as the sum of distances will generate a lesser value of Ω.

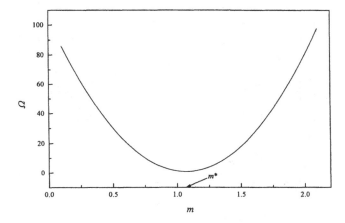

Figure 1.3. A simple search space, created from the data of Figure 1.1, Equation (1.3) and a large number of guesses of the value of m. This is an example of a minimisation problem, where the optimum is located at the lowest point.

This approach, of estimating an unknown parameter, or parameters, by simply solving the problem for a very large number of values of the unknowns is called an enumerative search. It is only really useful if there are relatively few unknown parameters and one can estimate Ω rapidly. As an example why such an approach can quickly run into problems of scale, consider the following. A problem in which there are ten unknowns, each of which are required to an accuracy of one percent, will require 100^{10}, or 1×10^{20}, estimations. If the computer can make 1000 estimations per second, then the answer will take over 3×10^9 years to emerge. Given that ten is not a very large number of unknowns, one percent not a very demanding level of accuracy and one thousand evaluations per second more than respectable for many problems, clearly there is a need to find a better approach.

Returning to Figure 1.3, a brief consideration of the shape of the curve suggests another approach: guess two possible values of m, labelled m_1 and m_2 (see Figure 1.4), then if $\Omega(m_1) > \Omega(m_2)$, make the next guess at some point m_3 where $m_3 = m_2 + \delta$, or else head the other way. Given some suitable, dynamic, way of adjusting the value of δ, the method will rapidly home in on m^*.

8

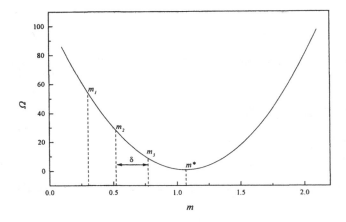

Figure 1.4. A simple, yet effective, method of locating m^*. δ is reduced as the minimum is approached.

Such an approach is described as a *direct* search (because it does not make use of derivatives or other information). The problem illustrated is one of minimisation. If $1/\Omega$ were plotted, the problem would have been transformed into one of maximisation and the desire would been to locate the top of the hill.

Unfortunately, such methods cannot be universally applied. Given a different problem, still with a single adjustable parameter, a, Ω might take the form shown in Figure 1.5.

If either the direct search algorithm outlined above or a simple calculus based approach is used, the final estimate of a will depend on where in the search space the algorithm was started. Making the initial guess at $a = a_2$, will indeed lead to the correct (or *global*) minimum, a^*. However, if $a = a_1$ is used then only a^{**} will be reached (a *local* minimum).

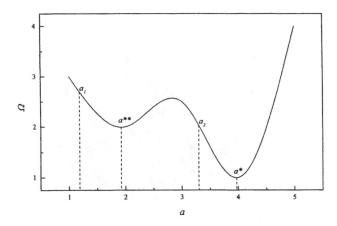

Figure 1.5. A more complex one-dimensional search space with both a global and a local minimum.

This highlights a serious problem. If the results produced by a search algorithm depend on the starting point, then there will be little confidence in the answers generated. In the case illustrated, one way around this problem would be to start the problem from a series of points and then assume that the true global minimum lies at the lowest minimum identified. This is a frequently adopted strategy. Unfortunately Figure 1.5 represents a very simple search space. In a more complex space (such as Figure 1.6) there may be very many local optima and the approach becomes unrealistic.

So, how are complex spaces to be tackled? Many possible approaches have been suggested and found favour, such as random searches and simulated annealing [DA87]. Some of the most successful and robust have proved to be random searches directed by analogies with natural selection and natural genetics—genetic algorithms.

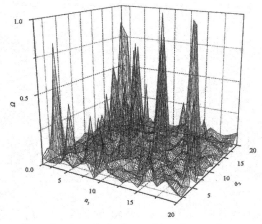

Figure 1.6. Even in a two-dimensional maximisation problem the search space can be highly complex.

1.3 GENETIC ALGORITHMS

Rather than starting from a single point (or guess) within the search space, GAs are initialised with a *population* of guesses. These are usually random and will be spread throughout the search space. A typical algorithm then uses three operators, *selection*, *crossover* and *mutation* (chosen in part by analogy with the natural world) to direct the population (over a series of time steps or *generations*) towards convergence at the global optimum.

Typically, these initial guesses are held as binary encodings (or *strings*) of the true variables, although an increasing number of GAs use "real-valued" (i.e. base-10) encodings, or encodings that have been chosen to mimic in some manner the natural data structure of the problem. This initial population is then processed by the three main operators.

Selection attempts to apply pressure upon the population in a manner similar to that of natural selection found in biological systems. Poorer performing individuals are weeded out and better performing, or *fitter*, individuals have a greater than average chance of promoting the information they contain within the next generation.

Crossover allows solutions to exchange information in a way similar to that used by a natural organism undergoing sexual reproduction. One method (termed *single point crossover*) is to choose pairs of individuals promoted by the selection operator, randomly choose a single locus (point) within the binary strings and swap all the information (digits) to the right of this locus between the two individuals.

Mutation is used to randomly change (flip) the value of single bits within individual strings. Mutation is typically used very sparingly.

After selection, crossover and mutation have been applied to the initial population, a new population will have been formed and the generational counter is increased by one. This process of selection, crossover and mutation is continued until a fixed number of generations have elapsed or some form of convergence criterion has been met.

On a first encounter it is far from obvious that this process is ever likely to discover the global optimum, let alone form the basis of a general and highly effective search algorithm. However, the application of the technique to numerous problems across a wide diversity of fields has shown that it does exactly this. The ultimate proof of the utility of the approach possibly lies with the demonstrated success of life on earth.

1.4 AN EXAMPLE

There are many things that have to be decided upon before applying a GA to a particular problem, including:

- the method of encoding the unknown parameters (as binary strings, base-10 numbers, etc.);
- how to exchange information contained between the strings or encodings;
- the population size—typical values are in the range 20 to 1000, but can be smaller or much larger;
- how to apply the concept of mutation to the representation; and
- the termination criterion.

Many papers have been written discussing the advantages of one encoding over another; or how, for a particular problem, the population size might be chosen [GO89b]; about the difference in performance of various exchange mechanisms and on whether mutation rates ought to be high or low. However, these papers have naturally concerned themselves with computer experiments, using a small number of simple test functions, and it is often not clear **how general** such results are. In reality the only way to proceed is to look at what others with similar problems have tried, then choose an approach that both seems sensible for the problem at hand and that you have confidence in being able to code up.

A trivial problem might be to maximise a function, $f(x)$, where

$$f(x) = x^2 \text{ ; for integer } x \text{ and } 0 \le x \le 4095.$$

There are of course other ways of finding the answer ($x = 4095$) to this problem than using a GA, but its simplicity makes it ideal as an example.

Firstly, the exact form of the algorithm must be decided upon. As mentioned earlier, GAs can take many forms. This allows a wealth of freedom in the details of the algorithm. The following (Algorithm 1) represents just one possibility.

1. Form a population, of eight random binary strings of length twelve (e.g. *101001101010, 110011001100, ...*).
2. Decode each binary string to an integer x (i.e. *000000000111* implies $x = 7$, *000000000000* = 0, *111111111111* = 4095).
3. Test these numbers as solutions to the problem $f(x) = x^2$ and assign a fitness to each individual equal to the value of $f(x)$ (e.g. the solution $x = 7$ has a fitness of $7^2 = 49$).
4. Select the best half (those with highest fitness) of the population to go forward to the next generation.
5. Pick pairs of *parent* strings at random (with each string being selected exactly once) from these more successful individuals to undergo single point crossover. Taking each pair in turn, choose a random point between the end points of the string, cut the strings at this point and exchange the tails, creating pairs of *child* strings.
6. Apply mutation to the children by occasionally (with probability one in six) flipping a *0* to a *1* or vice versa.
7. Allow these new strings, together with their parents to form the new population, which will still contain only eight members.
8. Return to Step 2, and repeat until fifty generations have elapsed.

Algorithm 1. A very simple genetic algorithm.

To further clarify the crossover operator, imagine two strings, *000100011100* and *111001101010*. Performing crossover between the third and fourth characters produces two new strings:

parents	children
000/100011100	*000001101010*
111/001101010	*111100011100*

It is this process of crossover which is responsible for much of the power of genetic algorithms.

Returning to the example, let the initial population be:

population member	string	x	fitness
1	*110101100100*	3428	11751184
2	*010100010111*	1303	1697809
3	*101111101110*	3054	9326916
4	*010100001100*	1292	1669264
5	*011101011101*	1885	3553225
6	*101101001001*	2889	8346321
7	*101011011010*	2778	7717284
8	*010011010101*	1237	1530169

Population members 1, 3, 6 and 7 have the highest fitness. Deleting those four with the least fitness provides a temporary reduced population ready to undergo crossover:

temp. pop. member	string	x	fitness
1	*110101100100*	3428	11751184
2	*101111101110*	3054	9326916
3	*101101001001*	2889	8346321
4	*101011011010*	2778	7717284

Pairs of strings are now chosen at random (each exactly once): 1 is paired with 2, 3 with 4. Selecting, again at random, a crossover point for each pair of strings (marked by a /), four new children are formed and the new population, consisting of parents and offspring only, becomes (note that mutation is being ignored at present):

population member	string	x	fitness
1	11/0101100100	3428	11751184
2	10/1111101110	3054	9326916
3	101101/001001	2889	8346321
4	101011/011010	2778	7717284
5	111111101110	4078	16630084
6	100101100100	2404	5779216
7	101101011010	2906	8444836
8	101011001001	2761	7623121

The initial population had an average fitness f_{ave} of 5065797 and the fittest individual had a fitness, f_{max}, of 11751184. In the second generation, these have risen to: f_{ave} = 8402107 and f_{max} = 16630084. The next temporary population becomes:

temp. pop. member	string	x	fitness
1	110101100100	3428	11751184
2	101111101110	3054	9326916
3	101101011010	2906	8444836
4	111111101110	4078	16630084

This temporary population does not contain a *1* as the last digit in any of the strings (whereas the initial population did). This implies that no string from this moment on can contain such a digit and the maximum value that can evolve will be *11111111110*—after which point this string will reproduce so as to dominate the population. This domination of the population by a single sub-optimal string gives a first indication of why mutation might be important. Any further populations will only contain the same, identical string. This is because the crossover operator can only swap bits between strings, not introduce any new information. Mutation can thus be seen in part as an operator charged with maintaining the genetic diversity of the population by preserving the diversity embodied in the initial generation. (For a discussion of the relative benefits of mutation and crossover, see [SP93a].)

The inclusion of mutation allows the population to leapfrog over this sticking point. It is worth reiterating that the initial population did include a *1* in **all** positions. Thus the mutation operator is not necessarily inventing new information but simply working as an insurance policy against premature loss of genetic information.

Rerunning the algorithm from the same initial population, but with mutation, allows the string *11111111111* to evolve and the global optimum to be found. The progress of the algorithm (starting with a different initial population), with and without mutation, as a function of generation is shown in Figure 1.7. Mutation has been included by visiting every bit in each new child string, throwing a random number between 0 and 1 and if this number is less than $^1/_{12}$, flipping the value of the bit.

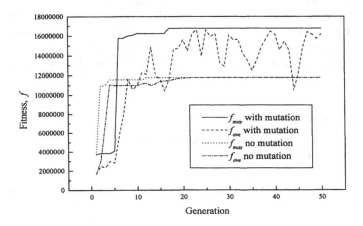

Figure 1.7. The evolution of the population. The fitness of the best performing individual, f_{max}, is seen to improve with generation as is the average fitness of the population, f_{ave}. Without mutation the lack of a *1* in all positions limits the final solution.

Although a genetic algorithm has now been successfully constructed and applied to a simple problem, it is obvious that many questions remain. In particular, how are problems with more than one unknown dealt with, and how are problems with real (or complex) valued parameters to be tackled? These and other questions are discussed in the next chapter.

1.5 SUMMARY

In this chapter genetic algorithms have been introduced as general search algorithms based on metaphors with natural selection and natural genetics. The central differences between the approach and more traditional algorithms are:

the manipulation of a population of solutions in parallel, rather than the sequential adjustment of a single solution; the use of encoded representations of the solutions, rather than the solutions themselves; and the use of a series of stochastic (i.e. random based) operators.

The approach has been shown to be successful over a growing range of difficult problems. Much of this proven utility arises from the way the population navigates its way around complex search spaces (or *fitness landscapes*) so as to avoid entrapment by local optima.

The three central operators behind the method are selection, crossover and mutation. Using these operators a very simple GA has been constructed (Algorithm 1) and applied to a trivial problem. In the next chapter these operators will be combined once more, but in a form capable of tackling more difficult problems.

1.6 EXERCISES

1. Given a string of length ten, what is the greatest value of an unknown Algorithm 1 can search for?

2. What is the resolution of Algorithm 1 when working with a string length of thirty?

3. Given a string length of 20 and a probability of mutation of $^1/_{20}$ per bit, what is the probability that a string will emerge from the mutation operator unscathed?

4. Implement Algorithm 1 on a computer and adapt it to find the value of x that maximises $\sin^4(x)$, $0 \leq x \leq \pi$ to an accuracy of at least one part in a million (Use a population size of fifty and a mutation rate of 1/(twice the string length).) This will require finding a transformation between the binary strings and x such that *000...000* implies $x = 0$ and *111...111* implies $x = \pi$.

5. Experiment with your program and the problem of Question 4 by estimating the average number of evaluations of $\sin^4(x)$ required to locate the maximum; (a) as a function of the population size, (b) with, and without, the use of crossover. (Use a mutation rate of 1/(twice the string length).)

CHAPTER 2

IMPROVING THE ALGORITHM

Although the example presented in Chapter 1 was useful, it left many questions unanswered. The most pressing of these are:

- How will the algorithm perform across a wider range of problems?
- How are non-integer unknowns tackled?
- How are problems of more than one unknown dealt with?
- Are there better ways to define the selection operator that distinguishes between good and very good solutions?

Following the approach taken by Goldberg [GO89], an attempt will be made to answer these questions by slowly developing the knowledge required to produce a practical genetic algorithm together with the necessary computer code. The algorithm and code go by the name *Little Genetic Algorithm* or LGA. Goldberg introduced an algorithm and PASCAL code called the *Simple Genetic Algorithm*, or SGA. LGA shares much in common with SGA, but also contains several differences. LGA is also similar to algorithms used by several other authors and researchers.

Before the first of the above questions can be answered, some of the terminology used in the chapter needs clarifying, and in particular, its relation to terms used in the life sciences.

2.1 COMPARISON OF BIOLOGICAL AND GA TERMINOLOGY

Much of the terminology used by the GA community is based, via analogy, on that used by biologists. The analogies are somewhat strained, but are still useful. The binary (or other) string can be considered to be a chromosome, and since only individuals with a single string are considered here, this chromosome is also the genotype. The organism, or phenotype, is then the result produced by the expression of the genotype within the environment. In

GAs this will be a particular set of unknown parameters, or an individual solution vector. These correspondences are summarised in Table 2.1.

Biological	GA
Chromosome or genotype	Structure, or string (often binary)
Locus	A particular (bit) position on the string
Phenotype	Parameter set or solution vector (real-valued)

Table 2.1 Comparison of biological and GA terminology.

2.2 ROBUSTNESS

Although the GA has, not unsurprisingly, proved itself able to find the maximum value of x^2 over a small range of integers, how is it likely to perform on a wider range of more realistic problems? This requires a consideration of what exactly is meant by *perform*.

The shortest and most efficient algorithm for generating the answer to a particular problem is simply a statement containing the answer to the problem. Given that this requires knowing the answer in the first place, the approach has little value. More useful are highly efficient methods that are specifically tailored to the application at hand, possibly containing problem specific operators and information. Such methods are likely to be efficient when working on the problem for which they were designed, but likely to be far less efficient—or even fail—when used on other problems. At the far end of the scale are robust techniques of almost universal application. Such techniques can, with little adaptation, work on a wide variety of problems but are likely to be much less efficient than highly tailored problem-specific algorithms. GAs are naturally robust algorithms, that by suitable adjustment of their operators and data encoding can also be made highly efficient. Given enough information about the search space it will always be possible to construct a search method that will outperform a GA. However, obtaining such information is for many problems almost as difficult as solving the problem itself. The "applicability" or *robustness* of the GA is illustrated in Figure 2.1: although highly problem-specific methods can outperform a GA, their domain of applicability is small. By suitable small adjustments to a GA, the algorithm can be made more efficient whilst still retaining a high degree of robustness.

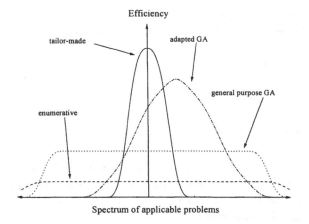

Figure 2.1. Comparison of the robustness of GA-based and more traditional methods. The more robust the algorithm the greater the range of problems it can be applied to. A tailor-made method such as a traditional calculus based algorithm might be highly efficient for some problems, but will fail on others. GAs are naturally robust and therefore effective across a wide range of problems.

2.3 NON-INTEGER UNKNOWNS

In Chapter 1 integer-valued parameters were represented as binary strings. This representation must now be adapted to allow for real-valued parameters. This requires providing a binary representation of numbers such as 2.39×10^{-6} or -4.91. (Another approach discussed later is the use of a real-valued representation within the GA, but this requires the redefinition of several of the GA operators.) There are many ways of doing this; however the most common is by a linear mapping between the real numbers and a binary representation of fixed length.

To carry out this transformation, the binary string (or genotype) is firstly converted to a base-10 integer, z. This integer is then transformed to a real number, r, using:

$$r = mz + c \ .$$

(2.1)

The values of m and c depend on the location and width of the search space. Expressions for m and c can be derived from the two simultaneous equations:

$$r_{min} = mz_{min} + c \qquad (2.2)$$

and

$$r_{max} = mz_{max} + c \qquad (2.3)$$

where r_{min}, r_{max}, z_{min} and z_{max} represent the minimum and maximum possible parameters in real and integer representations respectively. The smallest binary number that can be represented is of the form *000....0* which equates to 0 in base-10, so $z_{min} = 0$. z_{max} is given by:

$$z_{max} = 2^l - 1 \qquad (2.4)$$

where l is the length of the binary string used.

Subtracting (2.2) from (2.3) gives:

$$r_{max} - r_{min} = m(z_{max} - z_{min})$$

or

$$m = \frac{r_{max} - r_{min}}{z_{max} - z_{min}} .$$

Applying (2.4) and remembering that $z_{min} = 0$ gives:

$$m = \frac{r_{max} - r_{min}}{2^l - 1} . \qquad (2.5)$$

Finally, rearranging (2.2) gives:

$$c = r_{min} - mz_{min}$$

or (as $z_{min} = 0$)

$$c = r_{min} \ .$$

(2.6)

Equations (2.1), (2.5) and (2.6) then define the required transformation:

$$r = \frac{r_{max} - r_{min}}{2^l - 1} z + r_{min} \ .$$

(2.7)

AN EXAMPLE

Given a problem where the unknown parameter x being sought is known to lie between 2.2 and 3.9, the binary string *10101* is mapped to this space as follows:

$x = 10101$ therefore $z = 21$.

Using (2.7):

$$r = \frac{3.9 - 2.2}{2^5 - 1} 21 + 2.2 = 3.3516 \ .$$

A QUESTION OF ACCURACY

In the example above, *10101* was mapped to a real number between 2.2 and 3. The next binary number above *10101* is *10110* = 22, which, using (2.7) implies $r = 3.4065$. This identifies a problem: it is not possible to specify any number between 3.3516 and 3.4065.

This is a fundamental problem with this type of representation. The only way to improve accuracy is either to reduce the size of the search space, or to increase the length of the strings used to represent the unknowns. It is possible to use different representations that remove this problem [MI94]; however for most problems this proves unnecessary. By not making the search space larger than required and by choosing a suitable string length, the required accuracy can usually be maintained. ($l = 20$ implies an accuracy better than one part in a million.) For problems with a large number of unknowns it is important to use the smallest possible string length for each parameter. This requirement is discussed in more detail in the Chapter 6.

COMPLEX NUMBERS

Problems with complex-valued unknowns can be tackled by treating the real and imaginary parts as a pair of separate real parameters. Thus the number of unknowns is doubled.

2.4 MULTIPARAMETER PROBLEMS

Extending the representation to problems with more than one unknown proves to be particularly simple. The M unknowns are each represented as sub-strings of length l, These sub-strings are then concatenated (joined together) to form an individual population member of length L, where:

$$L = \sum_{j=1}^{M} l_j \ .$$

For example, given a problem with two unknowns a and b, then if $a = 10110$ and $b = 11000$ for one guess at the solution, then by concatenation, the genotype is $a \oplus b = 1011011000$.

At this point two things should be made clear: firstly, there is no need for the sub-strings used to represent a and b to be of the same length; this allows varying degrees of accuracy to be assigned to different parameters; this, in turn, can greatly speed the search. Secondly, it should be realised that, in general, the crossover cut point will not be between parameters but within a parameter. On first association with GAs this cutting of parameter strings into parts and gluing them back together seems most unlikely to lead to much more than a random search. Why such an approach might be effective is the subject of Chapter 3.

2.5 MUTATION

In the natural world, several processes can cause mutation, the simplest being an error during replication. (Rates for bacteria are approximately 2×10^{-3} per genome per generation [FU90, BA96,p19].) With a simple binary representation, mutation is particularly easy to implement. With each new generation the whole population is swept, with every bit position in every string visited and very occasionally a 1 is flipped to a 0 or vice versa. The probability of mutation, P_m is typically of the order 0.001, i.e. one bit in every thousand will be mutated. However, just like everything else about GAs, the correct setting for P_m will be problem dependent. (Many have used $P_m \approx 1/L$, others

[SC89a] $P_m \approx 1/(N\sqrt{L}$), where N is the population size). It is probably true that too low a rate is likely to be less disastrous than too high a rate for most problems.

Many other mutation operators have been suggested, some of which will be considered in later chapters. Some authors [e.g. DA91] carry out mutation by visiting each bit position, throwing at random a 0 or a 1, and replacing the existing bit with this new value. As there is a 50% probability that the pre-existing bit and the replacement one are identical, mutation will only occur at half the rate suggested by the value of P_m. It is important to know which method is being used when trying to duplicate and extend the work of others.

2.6 SELECTION

Thus far, the selection operator has been particularly simple: the best 50% are selected to reproduce and the rest thrown away. This is a practical method but not the most common. One reason for this is that although it allows the best to reproduce (and stops the worst); it makes no distinction between "good" and "very good". Also, rather than only allowing poor solutions to go forward to the next generation with a much lower probability, it simply annihilates them (much reducing the genetic diversity of the population). A more common selection operator is *fitness-proportional*, or *roulette wheel,* selection. With this approach the probability of selection is proportional to an individual's fitness. The analogy with a roulette wheel arises because one can imagine the whole population forming a roulette wheel with the size of any individual's slot proportional to its fitness. The wheel is then spun and the figurative "ball" thrown in. The probability of the ball coming to rest in any particular slot is proportional to the arc of the slot and thus to the fitness of the corresponding individual. The approach is illustrated in Figure 2.2 for a population of six individuals (a, b, c, d, e and f) of fitness 2.7, 4.5, 1.1, 3.2, 1.3 and 7.3 respectively.

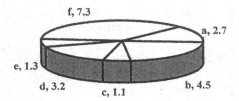

Figure 2.2. Roulette wheel selection. The greater an individual's fitness the larger its slot and the higher its chance of going forward to the next generation.

Implementing this operator is easier than it might seem. The circumference of the wheel is given by the sum of the population's fitness and the ball is represented by a random number between 0 and this sum. To discover which individual's slot the ball fell into, the fitness of the population members are added together one at a time, but this time stopping when this new sum exceeds the random number. At this point the last individual to be added is the selected individual. The algorithm is summarised in below:

1.	Sum the fitness of all the population members. Call this sum f_{sum}.
2.	Choose a random number, R_s, between 0 and f_{sum}.
3.	Add together the fitness of the population members (one at a time) stopping immediately when the sum is greater than R_s. The last individual added is the selected individual and a copy is passed to the next generation.

Algorithm 2. Implementing fitness-proportional selection.

The selection mechanism is applied twice (from Step 2) in order to select a pair of individuals to undergo, or not to undergo, crossover. Selection is continued until N (the population size, assumed to be even) individuals have been selected.

In this text the type of selection used is indicated by the value of φ, with $\varphi = \acute{r}$ indicating fitness-proportional (roulette wheel) selection.

2.7 ELITISM

Fitness-proportional selection does not guarantee the selection of any particular individual, including the fittest. Unless the fittest individual is much, much fitter than any other it will occasionally **not** be selected. To not be selected is to die. Thus with fitness-proportional selection the best solution to the problem discovered so far can be regularly thrown away. Although it appears counterproductive, this can be advantageous for some problems because it slows the algorithm, allowing it to explore more of the search space before convergence. This balance between the *exploration* of the search space and the *exploitation* of discoveries made within the space is a recurrent theme in GA theory. The more exploitation that is made the faster the progress of the algorithm, but the greater the possibility of the algorithm failing to finally locate the true global optimum. For many applications the search speed can be greatly improved by not losing the best, or *elite*, member between generations.

Ensuring the propagation of the elite member is termed *elitism* and requires that not only is the elite member selected, but a copy of it does not become disrupted by crossover or mutation.

In this text, the use of elitism is indicated by $\acute{e}$ (which can only take the value 0 or 1); if $\acute{e} = 1$ then elitism is being applied, if $\acute{e} = 0$ then elitism is not applied.

2.8 CROSSOVER

The Little Genetic Algorithm uses single point crossover as the recombination operator (in the natural world, between one and eight crossover points have been reported [GOT89,BA96,p18]). The pairs of individuals selected undergo crossover with probability P_c. A random number R_c is generated in the range 0-1, and the individuals undergo crossover if and only if $R_c \leq P_c$, otherwise the pair proceed without crossover. Typical values of P_c are 0.4 to 0.9. (If $P_c = 0.5$ then half the new population will be formed by selection and crossover, and half by selection alone.)

Without crossover, the average fitness of the population, f_{ave}, will climb until it equals the fitness of the fittest member, f_{max}. After this point it can only improve via mutation. Crossover provides a method whereby information for differing solutions can be melded to allow the exploration of new parts of the search space.

As described in Chapter 1, single point crossover proceeds by cutting the pair of selected strings at a random locus (picked by throwing a random number, R_L, between 1 and $L - 1$) and swapping the tails to create two child strings. For example, if $R_L = 4$, then:

Parents	Children
1010/0010101	*1010/1111111*
1111/1111111	*1111/0010101*

The new population now consists of N individuals (the same number as the original population) created by selection and crossover. Mutation then operates on the whole population except the elite member (if elitism is being applied). Once this is done, the old population is replaced by the new one and the generational counter, g, incremented by one.

2.9 INITIALISATION

Although as discussed in Chapter 1 the initial population is usually chosen at random, there are other possibilities. One possibility [BR91] is to carryout a series of initialisations for each individual and then pick the highest performing values. Alternatively, estimations can be made by other methods in an attempt to locate approximate solutions, and the algorithm can be started from such points.

2.10 THE LITTLE GENETIC ALGORITHM

Having now described how multi-parameter problems with non-integer unknowns can be tackled, and defined the mutation, selection, crossover and elitism operators, this knowledge can be brought together within a singular algorithm (Algorithm 3):

1. Generate an initial ($g = 1$) population of random binary strings of length $\sum_{k=1}^{M} l_k$, where M is the number of unknowns and l_k the length of binary string required by any unknown k. In general $l_k \neq l_j$; $k \neq j$.

2. Decode each individual, i, within the population to integers $z_{i,k}$ and then to real numbers $r_{i,k}$, to obtain the unknown parameters.

3. Test each individual in turn on the problem at hand and convert the objective function or performance, Ω_i, of each individual to a fitness f_i, where a better solution implies a higher fitness.

4. Select, by using fitness proportional selection, pairs of individuals and apply with probability P_c single point crossover. Repeat until a new temporary population of N individuals is formed.

5. Apply the mutation operator to every individual in the temporary population, by stepping bit-wise through each string, occasionally flipping a 0 to a 1 or vice versa. The probability of any bit mutating is given by P_m and is typically very small (for example, 0.001).

6. If elitism is required, and the temporary population does not contain a copy of an individual with at least the fitness of the elite member, replace (at random) one member of the temporary population with the elite member.

7. Replace the old population by the new temporary generation.

8. Increment, by 1, the generational counter (i.e. $g = g + 1$) and repeat from Step 2 until G generations have elapsed.

Algorithm 3. The Little Genetic Algorithm.

USING LGA

For many applications requiring the near optimisation of real or complex valued functions, LGA is a suitable algorithm. However, as mentioned several times already, the correct choice of algorithm is highly problem dependent and readers are encouraged to search the literature for successful applications of the technique to problems similar to their own. It may also prove worthwhile to consider some of the advanced operators discussed in Chapter 4. The application of GAs to help solve difficult problems has a long history and the number of adaptations to the basic technique is growing all the time. An algorithm as simple as LGA will not be suitable for all problems by any means.

28

INSTALLING AND RUNNING LGADOS

The disk enclosed with this book contains an implementation (LGADOS) of the LGA algorithm (although with $l_j = l_k$, for all j and k, i.e. all unknowns are represented with identical length strings). Both an uncompiled (LGADOS.BAS) and a compiled version (LGADOS.EXE) are included. The compiled version is designed for use with some of the exercises included at the end of chapters. The uncompiled version can be adapted for use with other problems.

A listing (in BASIC) of the program is given both on the disk and in Appendix B. BASIC has been used for several reasons. Firstly, it is one of the easiest computer languages to understand and should cause few problems for those with experience in FORTRAN, PASCAL or C. Secondly, it complements code written in PASCAL and C published in other introductory text books ([GO89] and [MI94] respectively). The disk also contains translations of the code into FORTRAN, PASCAL and C. To ensure maximum compatibility with the text, these are near direct translations from the BASIC code and therefore do not represent the best way of coding a GA in these languages. Those preferring to use a more professional FORTRAN GA should visit David Carroll's web site (see Appendix A) and download his GA code.

Updated versions of the programs will be found at http://www.ex.ac.uk/cee/ga/ . This site should be visited and any updates downloaded before the programs are used.

Lastly, most IBM compatible personal computers using Windows 3.x or DOS 6.x will have access to QBASIC—the environment used to write LGA—thereby allowing alterations to be made easily and the student exercises completed. However, QBASIC does not contain a compiler, and therefore LGADOS will run rather slowly on anything but the simplest problem. If you wish to adapt the code to use on your own problems you will probably need either to purchase a BASIC compiler (shareware ones are available), use one of the translations on the disk (LGADOS.F, LGADOS.PAS and LGADOS.C) or convert the code to a language for which you already have a compiler. This conversion should be relatively easy as no overly special structures or operators have been used.

A word of caution: LGADOS has been written with simplicity rather than efficiency in mind and as such does not represent good programming practice. There are much faster ways of performing some of the operations and better languages than BASIC in which to write such code. Those without the required knowledge may well find it advisable to enlist the help of a more experienced programmer to produce more efficient code. Having said this, in the majority of real-world problems to which GAs are applied, the time taken for the GA to cycle through a generation of selection, crossover and mutation is much less than the time taken to estimate the objective functions, and hence, the fitness of the individuals. This is a very different situation to that encountered when examining the performance of a GA on established test problems. The function estimations in such tests are typically much simpler and much quicker to calculate than their real-world counterparts—which might take several minutes per estimation. This implies that, in the majority of cases, time is better spent producing code that can carry out these function evaluations as rapidly as possible, rather than considering how to increase the speed of the GA's operators. BASIC allows—as do many other languages—the inclusion of mixed language routines. This means it is easy to add pre-existing routines to carry out the function evaluations.

Another sensible approach to developing applications is to use some of the GA code available on the World Wide Web (see Appendix A), or contact a researcher who has recently published in a closely related field using GAs and enthuse them into helping out.

LGADOS itself is examined in detail in Chapter 5, a quick glance at which shows that the program is quite short, and capable readers are strongly encouraged to write their own version, in the language of their choice. This is recommended even for those that may end up using someone else's code to solve research problems, as it ensures a full understanding of the processes involved.

To run the compiled version, copy LGADOS.EXE to a new sub-directory (folder) on your hard drive and type LGADOS from that directory if using DOS, or double-click on LGADOS.EXE from File Manager (or Explorer).

The user is presented with a series of options for N, P_c, P_m etc. and possible test problems (Figure 2.3).

Output is via two files: LGADOS.RES and LGADOS.ALL. LGADOS.RES lists the generation, g; fitness, f_{max}, of the highest performing individual; the average fitness of the generation, f_{ave}; and the unknown parameter values, r_{ki}, contained in the fittest individual. LGADOS.ALL lists g, f, r_k and the binary chromosome C for all individuals in all generations, and hence can be very large. The files are comma separated and can be loaded into most spreadsheets for analysis and data plotting.

```
N, Population Size (must be even) = 20
l, Substring Length (all sub-strings have the same length) = 10
G, Max. Generation = 50
Pc, Crossover Probability (>=0 and <=1) = 0.65
Pm, Mutation Probability (>=0 and <1) = 0.001
e, Elitism (1=on, 0=off) = 0
cm, Scaling Constant (a value of 0 implies no scaling) = 0
Problem (1=F1, 2=F2, 3=F3, 4=F4, 5=f^2) = 2
```

Figure 2.3. A completed input screen for LGADOS.EXE. The meaning of **cm** will described later, but it should be set to zero for now.

To test LGADOS, and the stochastic nature of GAs, a simple example:

$$MAX[f(x) = x^2] \; ; \; 0 \leq x \leq 10$$

can now be completed using LGADOS. The following settings should be used:

$N = 20$
$l = 10$
$G = 20$
$P_c = 0.6$
$P_m = 0.01$
$\acute{e} = 0$
$c_m = 0$
Problem $= f^2$

After setting these, press ENTER. LGADOS will display a simple listing of the average and best fitness within a single generation, together with the best estimate of x. These results are also stored within the file LGADOS.RES.

When LGA has run through all 20 generations print out LGADOS.RES, run the program again, and print LGADOS.RES. Finally repeat this process once more. If you compare the three sets of results they should be substantially different. (If not, you were simply fortunate.) The reasons for these differences are the stochastic processes embedded within the algorithm, i.e. the use of random numbers to pick individuals to mate, crossover sites and mutation locations. Therefore no two runs of a GA are ever likely to produce the same series of results (unless the random number generator was seeded with the same number both times). This is important to remember. If computer experiments are being run in an attempt to ascertain the best values of the internal GA settings (P_m, P_c etc.) for a particular problem, the results from one

GA run should not be relied upon to be meaningful. Rapid, or slow, progress of the GA could well be simply the result of the particular random numbers encountered (Figure 2.4).

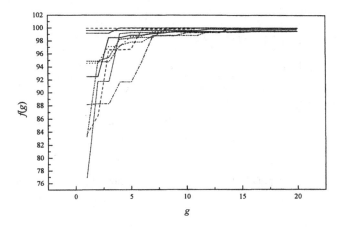

Figure 2.4. Sample results from multiple runs of LGADOS on the problem MAX[$f(x) = x^2$] ; 0 ≤ x ≤10. The runs show very different characteristics (f_{max} plotted).

It should be noted that although averaging solution vectors (i.e. parameter values) produced by GAs provides a way of monitoring progress and producing performance measures for the algorithm, there is little point in averaging solution vectors when dealing with real problems once the internal GA settings have been established. In fact, not only is averaging solution vectors of little benefit, but it can also lead to quite erroneous solutions. Figure 2.5 shows a hypothetical one-dimensional fitness landscape. If two runs of a GA produce the solutions a_1 and a_2 respectively, then the mean of these solutions is a_3—a very poor result. Such a space would be better tackled with a long GA run, whilst ensuring the population remained diverse, or by multiple runs disregarding all but the best solution.

In this text, the number of multiple runs made to produce a result is denoted by Я.

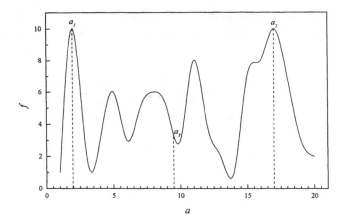

Figure 2.5. A potential pitfall of averaging solution vectors from multiple runs of a GA, $a_3 = (a_1 + a_2)/2$.

2.11 OTHER EVOLUTIONARY APPROACHES

Genetic algorithms are not the only evolutionary approach to search and optimisation.

Evolutionary Programming [FO66] typically uses a representation tailored to the problem (e.g. reals, not binary). All N individuals are selected and a representation specific, adaptive mutation operator used to produce N offspring. The next generation is then selected from the $2N$ individuals via a fitness-biased selection operator.

Evolution Strategies originally used $N = 1$ together with a mutation and selection operator. This has been extended [SC81] to $N > 1$, with mutation and recombination to create more than N offspring. Selection is then used to return the population to N individuals.

For an overview of such approaches see [BA96, p57-60, BA91 and SP93].

2.12 SUMMARY

In this chapter the algorithm has been extended to deal with multi-dimensional problems of non-integer unknowns. The selection operator has also been improved to allow use to be made of the distinction between "good" and "very good".

A comparison of biological and GA terminology has been made and the robustness of the algorithm qualitatively compared to more traditional methods.

The problem of limited accuracy caused by discretisation of the search space, implied by the use of a fixed binary representation, has been considered and seen to cause few difficulties for many problems.

A simple genetic algorithm, LGA, has been introduced and used to solve a trivial problem. This has allowed one of the potential pitfalls caused by the stochastic nature of the method to be discussed.

In the next chapter, some of the reasons why GAs work well across a wide range of problems will be further considered.

2.13 EXERCISES

1. Within the terminology used in GAs, characterise the difference between the genotype and the phenotype.

2. Derive (2.7)

3. Given $l = 4$, what is the highest fitness that can be found by a binary encoded GA for the problem MAX[$\sin^{10}(x)$]; $0 \leq x \leq 3$?

4. If $N = 6$ with $f_1 = 1$, $f_2 = 2$, $f_3 = 3$, $f_4 = 4$, $f_5 = 5$ and $f_6 = 6$, how many times is the individual with $f = 4$ likely to be picked by fitness proportional selection in a single generation? What is the minimum and maximum possible number of times the individual with $f = 6$ might be picked? What problem does this indicate could arise in a GA using fitness proportional selection?

5. Write, in a programming language of your choice, a GA similar to LGA.

6. Use LGADOS, or your own code, to study the evolution of a population whilst it explores the search space given by $f = \sum_{i=1}^{2} \sin^4(x_i)$; $0 \leq x_i \leq \pi$,

$i = 1, 2$. Plot the movements of the population across the search space (as a function of g) as a series of (x_1, x_2, f) surface plots.

CHAPTER 3

FOUNDATIONS

Although the roots of evolutionary inspired computing can be traced back to the earliest days of computer science, genetic algorithms themselves were invented in the 1960's by John Holland. His reasons for studying such systems went beyond a desire for a better search and optimisation algorithm. Such methods were (and still are) considered helpful abstractions for studying evolution itself in both natural and artificial settings. His book *Adaptation in Natural and Artificial Systems* from 1975 (and now updated) was, and still is, inspirational.

With the aid of his students Holland developed the GA further during the 1970's. He also produced a theoretical framework for understanding how GAs actually work. Until relatively recently this *schema theory* formed the basis of most theoretical work on the topic.

Exactly why Genetic Algorithms work is a subject of some controversy, with much more work being required before all questions are finally answered. However the subject is not without foundations. These foundations have emerged from two separate directions. One is based on attempts to provide a mathematical analysis of the underlying processes, the other on computer simulations on functions that reflect aspects of some of the problems to which GAs have been applied (or ones that GAs might have difficulty with).

There are some very good reasons why, even as practitioners rather than theorists, it might be beneficial for the subject to have a theoretical foundation. In particular, a knowledge of the type of problems where GAs might, or might not, perform well (or even work) would be extremely useful. Equally useful would be guidance on the degree to which such algorithms might outperform more traditional methods.

Much of the work in this area is not suitable for an introductory text. For an overview the reader should see the series *Foundations of Genetic Algorithms* [RA91,WH93,WH95]. However, a brief consideration of the subject is well worth the modest effort required. In the following, both a largely theoretical method and a more applied approach will be considered. The theoretical work is based on Holland's original schema theorem, popularised by Goldberg [GO89]. The applied work is based on the systematic adjustment of internal settings when using a GA to tackle a series of test functions.

Both approaches are required because while most theoretical work on GAs has concentrated on binary alphabets (i.e. strings containing only 0's and 1's), fitness-proportional selection and pseudoboolean functions (i.e. functions expressed using 0's and 1's), practitioners have used a vast array of representations and selection methods. Results therefore do not necessarily translate between these approaches, implying yet more caution when choosing settings and deciding between various algorithms etc.

3.1 HISTORICAL TEST FUNCTIONS

Before looking at schema theory there is a need to look at some of the theoretical test functions (or *artificial landscapes*) used to examine the performance of varying GAs. These functions are not only of historical interest. They, together with more complex functions, are often suited to the testing of user developed codes.

Although typical test functions are very useful because they allow for easy comparisons with other methods, they may have little relevance to real-world problems. Thus care must be taken not to jump to conclusions about what is best in the way of algorithm or settings. Often such functions have been too simple, too regular and of too low a dimension to represent real problems (see comments in [DA91b] and [WH95a]). Bäck [BA96,p138] suggests sets of functions should be used, with the group covering several important features. The set should:

1. consist exclusively of functions that are scalable with respect to their dimension M, i.e. the number of unknowns in the problem can be changed at will;

2. include a unimodal (i.e. single peaked), continuous function for comparison of convergence velocity (see below);

3. include a step function with several flat plateaux of different heights in order to test the behaviour of the algorithm in case of the absence of any local gradient information; and

4. cover multimodal (i.e. multi-peaked) functions of differing complexity.

Although many others had been investigating genetic algorithms for some time, De Jong's dissertation (published in 1975) *Analysis of the Behaviour of a Class of Genetic Adaptive Systems* [DE75] has proven to be a milestone. One reason for this is the way he carried out his computer experiments, carefully adjusting a single GA setting or operator at a time. The other is the range of functions (or problems) on which he chose to test the GA. These functions, together with additions, are still used today by some to make initial estimates of the performance of their own GAs. In fact it is well worth coding up a subset of these functions if you are writing your own GA (for function optimisation), simply so that you can check that all is proceeding according to plan within your program. If you are using a GA you did not write then this is still a worthwhile exercise to prove that you have understood the instructions. The idea of using test functions to probe the mechanics and performance of evolutionary algorithms has continued to the present day. For an excellent modern example see Bäck's recent book [BA96].

De Jong's suit of functions ranged from simple unimodal functions of few dimensions to highly multimodal functions in many dimensions. Unlike most research problems, all of them are very quickly calculated by computer and therefore many generations and experiments can be run in a short time. Adapted versions of three of the functions (together with some additions) are listed in Table 3.1, and two-dimensional versions of several of them presented in Figures 3.1a to 3.1d.

MEASURING PERFORMANCE

De Jong used two measures of the progress of the algorithm: the *off-line* performance and the *on-line* performance. The off-line performance f_{off} is a running average of the fitness of the best individual, f_{max}, in the population:

$$f_{off}(g) = \frac{1}{g} \sum_{j=1}^{g} f_{max}(j) \ .$$

The on-line performance, f_{on} is the average of all fitness values f_i calculated so far. It thus includes both good and bad guesses:

$$f_{on}(g) = \frac{1}{g} \sum_{j=1}^{g} \left[\frac{1}{N} \sum_{i=1}^{N} f_i(j) \right] .$$

Function	Limits
$f = F_1 = 79 - \sum_{j=1}^{3} x_j^2$	$-5.12 \le x_j \le 5.12$
$f = F_2 = 4000 - 100\left(x_1^2 - x_2\right)^2 + \left(1 - x_1\right)^2$	$-2.048 \le x_j \le 2.048$
$f = F_3 = 26 - \sum_{j=1}^{5} INT\left(x_j\right)$	$-5.12 \le x_j \le 5.12$
$f = F_4 = 0.5 - \dfrac{\left(\sin\sqrt{x_1^2 + x_2^2}\right)^2 - 0.5}{\left(1 + 0.001\left(x_1^2 + x_2^2\right)\right)^2}$	$-100 \le x_j \le 100$
$f = F_5 = A - \sum_{j=1}^{\le 30} \left[INT\left(x_j + 0.5\right)\right]^2$	$-40 \le x_j \le 60$
$f = F_6 = A - 20\exp\left[-0.2\sqrt{\dfrac{1}{m}\sum_{j=1}^{m(\le 30)} x_j^2}\right]$ $- \exp\left[\dfrac{1}{m}\sum_{j=1}^{m(\le 30)}\cos\left(2\pi x_j\right)\right] + 20$	$-20 \le x_j \le 30$

Table 3.1. Adapted versions of various test functions: De Jong's (F_1 to F_3), Davis (F_4) [DA91,SC89a] and Bäck (F_5 and F_6) [BA96]. The function $INT(-)$ returns the nearest integer less than or equal to (-). A is chosen to ensure a maximisation problem. Bäck [BA96] also presents an interesting fractal function based on the Weierstrass-Mandelbrot function [MA83,FE88].

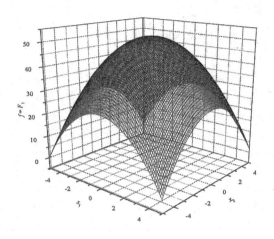

(a)

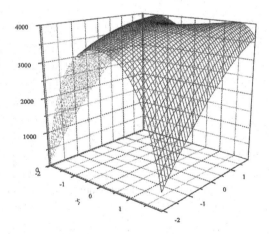

(b)

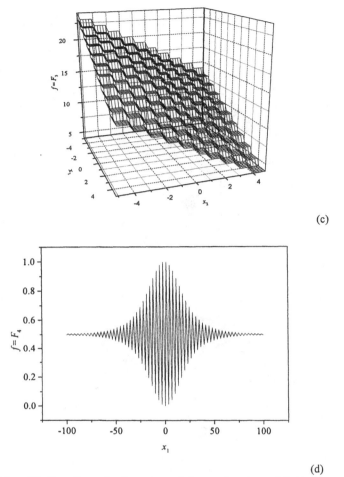

(c)

(d)

Figures 3.1 (a) to (d). Two dimensional versions of the test functions of Table 3.1: selected from [DE75 and GO89], F_1 to F_3; and a section through the global optimum of (F_4) [DA91,SC89a].

De Jong actually used six algorithms or *reproductive plans* for his GA experiments. Here, tests are restricted to testing the effect of mutation rate on the rate of convergence of LGA (studying the effect of N, P_c etc. is one of the exercises at the end of the chapter).

Another useful measure is the convergence velocity [adapted from BA96,p151], V:

$$V = \ln \sqrt{\frac{f^{max}(g = G)}{f^{max}(g = 0)}} \; .$$

It is important that such performance measures are averaged over $\mathcal{R}$ if sensible results are to be achieved. For complex multimodal functions, multiple runs are unlikely to find the same final optimum and one way of judging success is to plot a histogram of the number of times local optima of similar value were found.

Figures 3.2 and 3.3 show the effect of P_m on f_{max} (the maximum fitness in any generation) rather than f_{on} or f_{off}. In general, it is probably better practice to plot the number of objective function evaluations on the abscissa rather than the generation. This is because in tests where N varies or where f is not calculated for all individuals in each generation, this provides a better indication of computational effort. In fact, if $P_c \ll 1$ and $P_m \approx 0$ then very few new structures are created each generation, and LGA becomes a very wasteful algorithm because it re-calculates f_i for all i each generation.

The plots indicate that the success of different GA settings depends on the function. For F_1 $P_m = 0.3$ is better than $P_m = 0.003$, for F_3 the opposite is true. Further experimentation (left to the exercises at the end of the chapter) shows that similar results are obtained for the values of N and P_c. Interestingly, the independent estimation of the best value for each setting will not necessarily lead to the optimum set. In conclusion, there is no golden set of GA parameters; some sets work well with one type of function but less well with others. The same is true of the form of the algorithm itself.

42

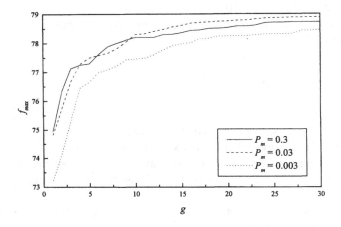

Figure 3.2. The progress of f_{max} for test function F_1 and various settings of P_m ($N = 20$, $P_c = 0.65$, $l = 10$, $\acute{e} = 1$, $\mathcal{A} = 20$).

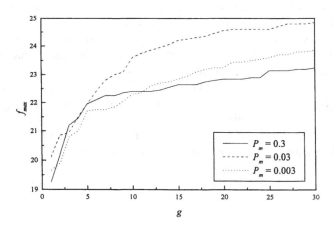

Figure 3.3. The progress of f_{max} for test function F_3 and various settings of P_m ($N = 20$, $P_c = 0.65$, $l = 10$, $\acute{e} = 1$, $\mathcal{A} = 20$).

THE PROBLEM OF CONVERGENCE

The above experiments indicate that, although LGA was able to find approximate optimum values of the test functions quickly, its progress was by no means linear. Initial progress is rapid and the value of f_{max} soars. However this progress is not maintained. One clue to the reason for this behaviour can be gleaned if the level of genetic diversity within the population is plotted against generation; another from a consideration of the likelihood of progression to the next generation under roulette wheel selection.

THE APPLICATION OF SCALING

If early during a run one particularly fit individual is produced, fitness proportional selection can allow a large number of copies to rapidly flood the subsequent generations. Although this will give rapid convergence, the convergence could quite possibly be erroneous or only to a local optimum. Furthermore, during the later stages of the simulation, when many of the individuals will be similar, fitness proportional selection will pick approximately equal numbers of individuals from the range of fitnesses present in the population. Thus there will be little pressure distinguishing between the good and the very good.

What is needed is a method whereby particularly good individuals can be stopped from running away with the population in the earlier stages, yet a degree of selection pressure maintained in the final stages. This can be achieved by various mechanisms; one being the use of *linear fitness scaling*.

Linear fitness scaling works by pivoting the fitness of the population members about the average population fitness. This allows an approximately constant proportion of copies of the best individuals to be selected compared with average individuals. Typical values for this constant, c_m, are in the range 1.0 to 2.0. When c_m equals 2, then approximately twice the number of best individuals will go forward to the next generation than will average individuals. To achieve this, the fitness of every population member will have to undergo a scaling just before selection. This scaling needs to be dynamic. The fitnesses will need to be drawn closer together during the initial stages and pulled further apart during the later generations. The required scaling is achieved using the linear transformation:

$$f_i^s(g) = a(g)f_i(g) + b(g) \;,$$

where f_i is the *true* fitness of an individual, i, and f_i^s the *scaled* fitness.

As already stated, the mean fitness of the population f_{ave} is assumed to remain unchanged, so:

$$f^s_{ave}(g) = f_{ave}(g) \ .$$

An additional requirement is that

$$f^s_{max}(g) = c_m(g) f_{ave}(g) \ ,$$

Where f^s_{max} is the scaled fitness of the best individual.

This implies that:

$$a(g) = \frac{(c_m - 1) f_{ave}(g)}{f_{max}(g) - f_{ave}(g)}$$

and

$$b(g) = (1 - a(g)) f_{ave}(g) \ .$$

Unfortunately, such a transformation can produce negative scaled fitnesses. These can be eliminated in various ways, the simplest (but rather crude) way being just setting any that occur to zero. In LGADOS, setting c_m itself to zero stops scaling from being applied.

Many other scaling procedures are possible and are discussed in the literature. An alternative approach is to use a different selection mechanism, as considered in Chapter 4.

Scaling can be important with even simple problems, as the following illustrates. LGA can be used to find the value of x which maximises $F = 1000 + \sin(x)$, $0 \leq x \leq \pi$; but as Figure 3.4 shows, there will be very little selection pressure on individuals as they will all have near identical performance. This implies that the algorithm will proceed rather slowly. Fitness scaling, as described above, could be used to increase the selection probabilities of better individuals. However, a simpler approach would be via a simple non-dynamic direct fitness function adaptation to ensure f spans a sensible range i.e. $f = F - 1000$. An alternative approach would be to use one of the selection mechanisms discussed in the next chapter.

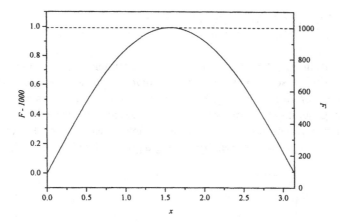

Figure 3.4. Increasing the difference between population members via a simple non-dynamic direct fitness function adaptation; F = dashed line, $F - 1000$ = solid line.

GENETIC DRIFT

The amount of diversity can be measured in several ways. An easily calculable measure is η_{max}, the genotypic similarity between the string representing the fittest individual and all the other members of the population.

To calculate η_{max} the value of each bit in the fittest string is compared with the value of the same bit in all the other strings in turn. Any matching bits increment η_{max} by 1. When all the positions have been compared the result is normalised by dividing by the total number of bits in the other strings, i.e. the product $(N-1)L$.

For example, given a population of four chromosomes of length five:

C_1 *10110*
C_2 *01111*
C_3 *10110*
C_4 *11110*

with C_1 having the highest fitness, then η_{max} is given by

$$\frac{2+1+3+3+2}{(4-1)\text{x}5} = \frac{11}{15} = 0.73 \ .$$

Plotting η_{max} for one of the earlier experiments gives Figure 3.5; here the population is seen to rapidly lose its diversity if scaling is not used. By including linear fitness scaling the diversity is seen to fall less rapidly in the first few generations, implying a greater degree of exploration. In later generations, η_{max} continues to rise in an almost linear fashion because of the higher selection pressure present (implying a greater degree of exploitation).

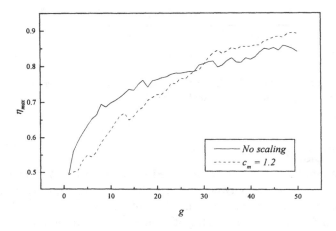

Figure 3.5. The progress of the similarity measure η_{max}. The use of scaling produces a more linear growth in η_{max} ($f = x^5$, $0 \leq x \leq \pi$, $N = 20$, $P_c = 0.65$, $P_m = 0.001$, $l = 10$, $\dot{\varepsilon} = 0$, $c_m = 0$ and 1.2, $\mathcal{A} = 20$).

3.2 SCHEMA THEORY

This is an approach introduced by Holland [HO75] and popularised by Goldberg [GO89].

A schema (plural schemata) is a fixed template describing a subset of strings with similarities at certain defined positions. Thus, strings which contain the same schema contain, to some degree, similar information. In keeping with the rest of this book, only binary alphabets will be considered, allowing templates to be represented by the ternary alphabet $\{0,1,\#\}$. Within any string the presence of the meta-symbol # at a position implies that either a *0* or a *1* could be present at that position. So for example,

101001

and

111001

are both instances of the schema

1##001.

Conversely, two examples of schemata that are contained within

010111

are

01#111

and

#101##.

Schemata are a useful conceptual tool for several reasons, one being that they are simply a notational convenience. Imagine a simple one-dimensional problem:

$$MAX\left[f(x) = x^2\right] ; 0 \le x \le 511$$

Clearly *f*(x) is at a maximum when *x* is maximum i.e. when *x* = 511 (Figure 3.6).

48

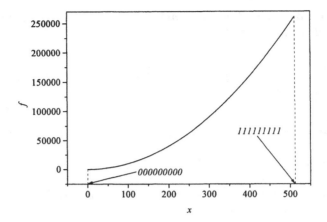

Figure 3.6. The one dimensional function $f(x) = x^2$.

511 is *111111111* in binary. Examples of binary numbers approximately equal to 511 are:

110101011,
111110100,
110001110 and
111011101.

Examples very far from 511 are:

000011101,
000000000,
000001010 and
000010010.

Contrasting these two sets of binary numbers it becomes apparent that the near maximum values of x are all instances of

11########.

Thus *11#######* provides a convenient notation for the numerous binary strings that represent a near optimal solution to the problem (Figure 3.7).

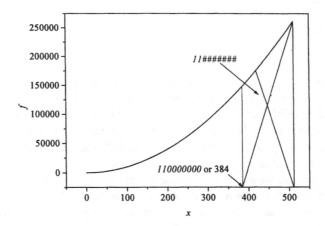

Figure 3.7. $f(x) = x^2$. The line $x = 384 = 110000000$, indicates the minimum value any string containing the schema *11#######* can have and is therefore the *boundary* of the schema. All other instances of this schema lie to the right of this line.

There are a great number of possible schemata within a single string. Given the randomly generated four bit string,

1011

the list of all possible schemata is

101#,
10#1,
1#11,
#011,
10##,
1#1#,
1##1,
#01#,
#0#1,
##11,

1###,

#0##,

##1#,

###1,

and

1011,

or 16 entries. $16 = 2^4$; for any real string of length L there are 2^L possible schemata. For an arbitrary string, each bit can take the value *1* or *0* or *#*. So there are 3 possibilities for each bit, or 3x3x3x3 possibilities for a string of length 4. For a string of length 200 (a number typical of that found in many real GA applications) there are therefore 3^{200} ($\approx$3x10^{95}) schemata to be found (c.f. 10^{80}, the number of stable particles in the universe).

In general, for an alphabet of cardinality (or distinct characters) k, there are $(k + 1)^L$ schemata. For a population of N real strings there are Nk^L possible schemata. The actual number of schemata within a population is likely to be much less than this for two reasons. Firstly, some schemata can simultaneously represent differing strings, for example, given $N = 2$, $L = 3$ and the population {*101, 111*}, a table of possible schemata can easily be formed (Table 3.2).

$C_1 = 101$	$C_2 = 111$
#01	*#11*
1#1	*1#1*
10#	*11#*
##1	*##1*
#0#	*#1#*
1##	*1##*
###	*###*
101	*111*

Table 3.2. Possible schemata for a particular population of 2 strings and $L = 3$.

This table contains 16 schemata (8 for each string), but only 8 are unique. For other populations this reduction may be less dramatic. If, once more, $N = 2$ and $l = 3$, a possible population is {*111,000*}; then there will be only one shared schema, namely *###* and hence there are 15 unique schemata. Secondly, not all the population members themselves are likely to be unique, particularly in an algorithm that has cycled through many generations and is near convergence. Thus the number of schemata in the population will change as the generations go by, but will always be < Nk^L.

Not all schemata are equal. The area of the search space represented by a schema and the location of this area depend on the number and location of the meta-symbols within the schema. Schema such as *1####* and *0####* include much larger regions of the search space than *1011#* or *0010#*. Schemata are typically classified by their *defining length* and their *order*. The order o of a schema S is the number of positions within the schema not defined by a meta-symbol, i.e.

$$o(S) = L - m ,$$

where m is the number of meta-symbols present within a string of length l. The order is therefore equal to the number of fixed positions within the schema:

$S = $ *#1#0#*; $o(S) = 2$
$S = $ *1101#*; $o(S) = 4$
$S = $ *11001*; $o(S) = 5$
$S = $ *#####*; $o(S) = 0$

The defining length d specifies the distance between the first and last non meta-symbol characters within the string:

$S = $ *#1#0#*; $d(S) = 4 - 2 = 2$
$S = $ *1101#*; $d(S) = 4 - 1 = 3$
$S = $ *11001*; $d(S) = 5 - 1 = 4$
$S = $ *#####*; $d(S) = 0 - 0 = 0$

It is worthwhile trying to visualise the different regions that such schemata cover. Figures 3.8 to 3.10 show the regions covered by a series of schemata for a one-dimensional problem. Most problems tackled using genetic algorithms have many more dimensions than this, but such a space makes it possible to get a feel for how schemata translate to physical regions within the problem space. In particular, low order schemata cover large regions of space and high-order schemata much smaller regions.

The defining length and the order of a schema are not the whole story. Two schemata with identical values of o and d can ring-fence very different regions of space. Both *##0#0* and *1#1##* have $o = 2$ and $d = 2$ but no overlap (Figures 3.9 and 3.10). Despite this, the order and the defining length of a schema are very important indicators of usefulness and chance of survival within a GA.

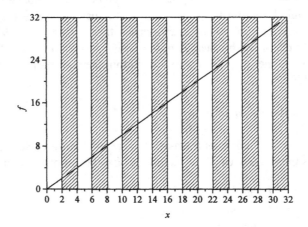

Figure 3.8. Sketch of ###1# and the simple function $f = x$; $0 \leq x \leq 31$.

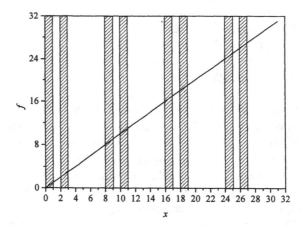

Figure 3.9. Sketch of ##0#0 and the simple function $f = x$; $0 \leq x \leq 31$.

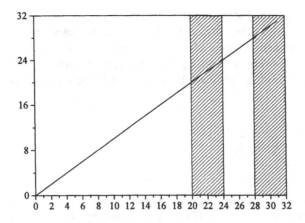

Figure 3.10. Sketch of *1#1##* and the simple function $f = x$; $0 \leq x \leq 31$.

3.3 SCHEMA PROCESSING

A lot of rather brutal things can happen to any particular string within a GA. It can get fragmented by crossover, attacked by mutation or simply thrown away by the selection operator. Despite all this violence, it is relatively easy to estimate how the number of instances of a particular schema might change during a GA run. This estimation throws light on just why GAs can be so successfully employed as optimisation routines, and why they might experience difficulty solving some types of problems. The result can only be an estimate because of the stochastic nature of the algorithm. The calculation proceeds as follows:

If Φ is the number of instances of any particular schema S within the population at generation g, such that:

$$\Phi(S,g) > 0$$

then it would seem reasonable that if on average Φ represents strings of above average fitness, then:

$$\overline{\Phi}(S, g+1) > \Phi(S,g)$$

where the bar over Φ indicates an estimate.

Conversely, schemata that, on average, represent poorly performing strings will see their numbers decline. More precisely, if selection is carried out on a fitness-proportional basis (e.g. by roulette wheel selection) the probability P_i of selection (during a single selection event) for any individual (or string) is given by:

$$P_i(g) = \frac{f_i(g)}{\sum_{j=1}^{N} f_j(g)} \ .$$

In such a system, schemata will feel this selection pressure via the strings that are instances of each schema. If $u(S,g)$ is the average fitness of all instances of S then:

$$\overline{\Phi}(S, g+1) = \frac{u(S,g)}{f_{ave}(g)} \Phi(S,g) \ . \tag{3.1}$$

Equation (3.1) is the *schema growth equation* (ignoring the effects of crossover and mutation) and shows that the number of instances of any schema S in the next generation depends on the value of u in the current generation.

Assuming that a particular schema remains constantly above average by βf_{ave}, (3.1) becomes:

$$\overline{\Phi}(S, g+1) = \frac{f_{ave}(g) + \beta f_{ave}(g)}{f_{ave}(g)} \Phi(S,g)$$

which is,

$$\overline{\Phi}(S, g+1) = (1 + \beta)\Phi(S,g)$$

or,

$$\overline{\Phi}(S, g) = (1 + \beta)^g \Phi(S, g = 0) \ .$$

This implies that better (worse) performing schemata will receive exponentially increasing (decreasing) numbers of trials in the next generation. There are only N acts of selection per generation (assuming the whole generation is replaced each generation) yet the algorithm manages to simultaneously allocate exponentially increasing (decreasing) numbers of trials to a vast number ($>> N$) of schemata, seemingly effortlessly.

THE EFFECT OF CROSSOVER

The degree of disruption caused to individual schemata by crossover will depend on the schemata involved. The chromosome

$$C_1 = 000011100$$

contains (amongst many) the following two schemata:

$$S_1 = 0########0$$

and

$$S_2 = ####11###$$

If C_1 is selected to mate with another string C_2, then if the cut point is between the fourth and fifth digits, i.e.

$$C_1 = 0000/11100$$

then S_2 will survive in at least one of the children, no matter the form of C_2. S_1 however will only survive if C_2 also contains identical bit values in the two fixed outer positions. For most possible cut points it is clear that S_2 will survive and S_1 will not. This is because $d_1 >> d_2$. More specifically, the probability of a schema being destroyed by crossover is less than:

$$\frac{d(S)}{L-1}.$$

Therefore, given a crossover probability of P_c, the chance of survival to the next generation is greater than or equal to:

$$1 - P_c \frac{d(S)}{L-1} \; .$$

Applying this reduction to the schema growth equation gives:

$$\overline{\Phi}(S, g+1) = \frac{u(S,g)}{f^{ave}(g)} \Phi(S,g) \left(1 - P_c \frac{d(S)}{L-1}\right) \; .$$

THE EFFECT OF MUTATION

The probability of a single bit surviving a single mutation is simply:

$$1 - P_m \; .$$

The greater the order of the schema the greater the probability of disruption. With $o(S)$ bits defined, the probability of the whole schema surviving will be:

$$\left(1 - P_m\right)^{o(S)} .$$

Applying this in turn to the schema growth equation, and ignoring lesser terms, gives:

$$\boxed{\overline{\Phi}(S, g+1) = \frac{u(S,g)}{f^{ave}(g)} \Phi(S,g) \left(1 - P_c \frac{d(S)}{L-1} - o(S)P_m\right)}$$

Thus <u>short</u>, <u>low-order</u>, <u>above-average</u> schemata are given exponentially increasing numbers of trials in subsequent generations. Such schemata are termed *building blocks*. The *building block hypothesis* states that GAs attempt to find highly fit solutions to the problem at hand by the juxtaposition of these building bocks [MI94, p51] (see [FO93] and [AL95] for criticisms).

Somewhere between 2^L and $N2^L$ schemata are being processed by the GA each generation. Many will be disrupted by mutation and crossover but it is possible (using arguments which lie outside the scope of this book) to estimate a lower bound on the number that are being processed usefully, i.e. at an exponentially increasing rate. The answer is of order N^3 (see [BE93] for a recent discussion on this). The ability of the GA to process N^3 schemata each generation while only processing N structures has been given the name *implicit parallelism* [HO75,GR91,GR89].

DECEPTION

The above indicates that the algorithm might experience problems where it is possible for some building blocks to *deceive* the GA and thereby to guide it to poor solutions, rather than good ones. For example, if f^* occurs at:

$C^* = 000000$, and

$S_1 = 00\#\#\#\#$ and
$S_2 = \#\#\#\#00$

represent (<u>on average</u>) above average solutions, then convergence would seem guaranteed. However, if the combination of S_1 and S_2:

$S_3 = 00\#\#00$

is (<u>on average</u>) very poor, then the construction of C^* might cause difficulties for the algorithm.

Deception in GAs shares similarities with epistasis in biological systems, where the existence of a particular gene affects genes at other loci. With sufficient knowledge of the problem at hand it should be possible to always construct encodings such that deception is avoided. However, for many real-world problems this task might be of similar complexity to that of solving the problem itself. (See [GR93] for a discussion).

3.4 OTHER THEORETICAL APPROACHES

Although schema analysis indicates that, via the exponential allocation of trials, the GA might form the basis of a highly efficient search algorithm, it leaves many questions unanswered. There has been much debate on how relevant the approach is for GAs working with real problems [MI96, p125-126]. Others [MI92,FO93,MI94b] have concentrated on the role of crossover rather than selection. It is interesting to note that although the analysis indicates the use of minimal alphabets (i.e. binary) because they offer the greatest number of schemata, and the use of fitness proportional selection, those working with real-world problems have found other encodings and selection mechanisms to be superior [WH89].

For an excellent introduction to some of these, and other, ideas the reader is directed to reference [MI96, p125-152] and the discussions in [AN89].

3.5 Summary

In this chapter, the idea of using test functions to ensure the correct operation of the algorithm has been introduced. Such functions can also be used to study the effect of various settings within the GA. The subject of premature convergence has been seen to be controllable to some extent by finding the correct balance between exploration and exploitation. A useful technique to help with this has been found to be the use of scaling.

Schema theory has been outlined and the schema growth equation stated. The relative proportions of the search space covered by schemata of various orders and defining lengths has been demonstrated.

There is much to do in the field of the mathematical foundations of GAs. In particular, there is a need to discover how performance can be maximised for various problem classes and how deception can be identified and avoided.

In the next chapter combinatorial optimisation will be considered, along with the rather strange ideas of niches and species. Multicriteria problems are then discussed along with several advanced operators.

3.6 Exercises

1. Use LGADOS.EXE, or your own GA, to study the effect of N, L, P_c, P_m and $\acute{e}$ on performance when maximising test functions F_1 to F_4 of Table 3.1. (LGADOS.EXE numbers the functions in the same manner as Table 1.)

2. Try and identify a <u>single</u> optimal set of N, P_c, P_m and $\acute{e}$ for the test functions of Table 3.1.

3. Add linear fitness scaling and the ability to calculate η_{max} to your GA.

4. Use LGADOS.EXE, or your own GA, to study the effect of c_m on performance and η_{max} when maximising the test functions of Table 3.1.

5. For $f = x^2$, $0 \leq x \leq 127$, what is the average fitness of $1\#\#\#\#\#\#$, $0\#\#\#\#\#\#$ and $11\#\#\#\#\#$?

6. Plot the progression with generation of the number of unique schemata within a GA for a simple problem. (A very low choice for L is recommended).

CHAPTER 4

ADVANCED OPERATORS

Although simple genetic algorithms, such as LGA, can be used to solve some problems, there are numerous extensions to the algorithm which have been developed to help improve performance and extend the range of applicability. These include more efficient crossover and selection methods, algorithms that deliberately hunt for local optima, combinatorial optimisation techniques, techniques to deal with multicriteria problems, and hybrid algorithms which combine the speed of traditional search techniques with the robustness of GAs.

Being an introductory text, many of these extensions can not be given full justice and can only be outlined. For additional details, readers are directed to other texts, in particular references [MI94, ZA97 and BA96]. In addition, some of the techniques are described in more detail and then applied to various problems in Chapter 6.

4.1 COMBINATORIAL OPTIMISATION

Many problems do not require the optimisation of a series of real-valued parameters, but the discovery of an ideal ordered list, the classic example being the Travelling Salesman Problem (TSP). In the TSP a fictitious salesperson is imagined having to find the shortest route, or tour, between a series of cities. Typically the rules state that no city is visited more than once. Other examples of such combinatorial problems are gas and water pipe placement, structural design, job-shop scheduling, and time tabling.

A great deal of effort has been applied to trying to find efficient algorithms for solving such problems and this work has continued with the introduction of GAs. The main problem with applying a genetic algorithm, as described so far, to such a problem is that crossover and mutation have the potential to create unfeasible tours. To see why this is, consider the TSP described by Figure 4.1. Here there are eight cities, labelled *a* to *h*, arranged randomly on a plane. Table 4.1 lists the relative distances between each city.

60

Figure 4.1. A TSP. What is the shortest tour that connects all eight cities?

City	a	b	c	d	e	f	g	h
a	0	17	27	73	61	57	51	23
b	17	0	37	73	72	74	66	40
c	27	37	0	48	35	49	65	50
d	73	73	48	0	47	82	113	95
e	61	72	35	47	0	38	80	78
f	57	74	49	82	38	0	48	65
g	51	66	65	113	80	48	0	40
h	23	40	50	95	78	65	40	0

Table 4.1. Distances in km between the cities in Figure 4.1.

One example of a tour might be (Figure 4.2),

b c d e g h f a,

another (Figure 4.3),

c b g f a d e h.

If single-point crossover is applied directly to these tours the result, cutting at the mid-point, is (Figure 4.4):

b c d e a d e f

and (Figure 4.5)

c b g f g h f a.

Neither of these represents a complete tour.

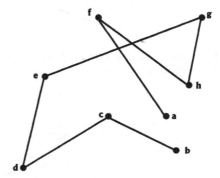

Figure 4.2. The tour *b c d e g h f a.*

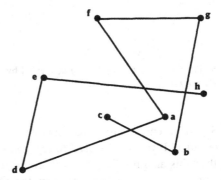

Figure 4.3. The tour *c b g f a d e h.*

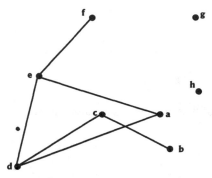

Figure 4.4. The partial tour *b c d e a d e f*, created by applying single point crossover to the tours *b c d e g h f a* and *c b g f a d e h*.

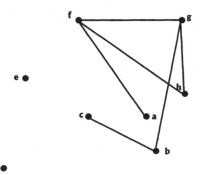

Figure 4.5. The partial tour *c b g f g h f a*, created by applying single point crossover to the tours *b c d e g h f a* and *c b g f a d e h*.

So, how can a crossover operator be designed that only generates complete tours? If the strings used to represent the tours are to remain of fixed length, then this also implies that each city can only be visited once. There are many ways of constructing such an operator. One would be to apply crossover as before, then reject any incomplete tours generated. However, this would require rejecting most tours and it is relatively easy to imagine far less wasteful algorithms. One possibility is *Partially Matched Crossover* (PMX) [GO89, p166-179].

The TSP representation described above gives the strings a very different property to the strings used to respect real-valued optimisation problems. The position and value of an element are not unrelated. In fact, within the TSP it is only the order which matters. Ideally, the new crossover and mutation operators must not only create feasible tours but also be able to combine building blocks from parents of above average fitness to produce even fitter tours.

PMX proceeds in a simple manner: parents are selected as before; *two* crossover sites are chosen at random (defining the *matching section*) then exchange operators are applied to build the two new child strings.

Returning to the previously defined tours, and selecting two cut points at random:

$$tour\ 1 = b\ c\ /\ d\ e\ g\ /\ h\ f\ a$$
$$tour\ 2 = c\ b\ /\ g\ f\ a\ /\ d\ e\ h.$$

First the whole centre portion or matching section is swapped:

$$tour\ 1' = b\ c\ /g\ f\ a\ /\ h\ f\ a$$
$$tour\ 2' = c\ b\ /\ d\ e\ g\ /\ d\ e\ h.$$

Neither of these tours is a feasible tour. In tour 1' there is no d or e and in tour 2' cities a and f are not visited. As the strings are of fixed length, this means that both tours visit some cities more than once. In the case of tour 1' cities a and f are visited twice, tour 2' visits d and e twice. By definition, one of these repeats is within the matching section and one without. Also by definition, any city that is visited twice by one tour must be missing from the other tour. This suggests a way forward. Cities that are visited twice in one tour are swapped with cities that are visited twice in the other tour. Only one representative (the one not in the matching section) of such cities is swapped—otherwise the process would be circular and unconstructive. So, in this example, a outside the matching section of tour 1' swaps with the d of tour 2', and similarly for the cities f and the e. The two tours:

$$tour\ 1'' = b\ c\ g\ f\ a\ h\ e\ d$$
$$tour\ 2'' = c\ b\ d\ e\ g\ a\ f\ h$$

are formed, each of which is complete.

PMX is relatively easy to implement within LGA by making suitable alterations to the crossover operator and setting P_m to zero.

4.2 LOCATING ALTERNATIVE SOLUTIONS USING NICHES AND SPECIES

In most optimisation problems the hunt is for the best possible solution. This might be the global optimum if this can be found, or a point in the vicinity of the global optimum if the problem is very large and difficult. However some problems are characterised by a search for a series of options rather than a unique solution vector. Problems in which these options reside at some distance from the global optimum are particularly interesting. In such cases there is a likelihood that the options are separated by regions which equate to much poorer solutions; rather than trying to avoid local optima, the idea is to try and hunt them down. Such a fitness landscape is illustrated by the multimodal function $f(x)$ shown in Figure 4.6. Although intuitively there is something distinctive about the values of x which equate with peaks in $f(x)$, mathematically none of them gives rise to a fitness greater than many of the points near the global optima. This begs the question, why hunt specifically for such values of x if any point near the global optimum is likely to generate higher fitness anyway?

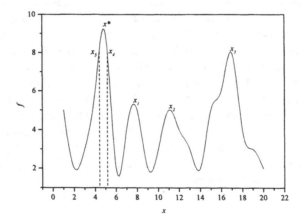

Figure 4.6. A multimodal function with a global optimum at x^* and secondary, or local optima, at x_1, x_2 and x_3. Any value of x between $x = x_4$ and $x = x_5$ will give rise to a fitness, f, greater than any of the local optima. So why even attempt to look for such local optima?

One reason for attempting such searches can be best explained by an example. If the problem characterised by the fitness landscape shown in Figure 4.6 was an architectural one, in which x was the angle of a pitched roof and f the inverse of the financial cost of the construction, then each local optima take on a significant meaning. They represent good, but not ideal, financial solutions of radically different form. If cost is the only criterion, then angle $x*$ is the only choice; however if any of the solutions x_1, x_2 or x_3 are deemed to be more in keeping with other, visual, desires then the architect might be able to convince the client to invest the small amount of extra capital required. Although there are many points close to the global optimum that offer better values of f than any of the local optima, their closeness to the global optimum may produce little justification for adopting such a design rather than the optimum. This is not so for those structures represented by the local optima.

In essence, the optimiser is being used as a filter, a filter charged with the task of finding highly performing, novel solutions to the problem across the whole problem space and ignoring, as much as possible, all other points.

One way of finding such optima is simply by the use of multiple runs. Beasley et. al. [BE93a,MI94,p176] indicate that if all optima have equal likelihood of being discovered then $\mathcal{R}$ should be set by:

$$\mathcal{R} \approx \rho(0.577 + \log \rho)$$

where ρ is the number of optima. As all optima will not generally be equally likely to be discovered, $\mathcal{R}$ will typically need to be much greater than this.

Figure 4.7 shows a more complex search space. The quest is for a technique that can effectively filter out any points that generate a fitness less than some minimum f_{min}. Such a filter would, if perfect, generate Figure 4.8, which is much easier to interpret than the original function. In the figures shown, this filtering is easy to apply by hand because the value of f is known at every point (x,y) within the problem space. Ordinarily f is likely to be far too complex to be estimated at more than a small fraction of the search space. So a technique is needed that can hunt for peaks within a complex landscape (producing Figure 4.9). This is somewhat ironic; up until now the central consideration has been the avoidance of local optima, now the desire is to actively seek them.

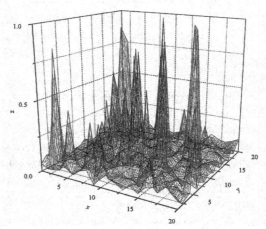

Figure 4.7. A more complex multi-peaked landscape; points of interest are those with $f > f_{min}$.

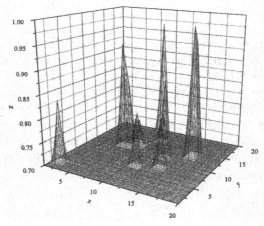

Figure 4.8. A filtered version of Figure 4.7 showing only points with $f > f_{min}$. The task of the GA is to approximate this filtering without estimating f at all points (x,y).

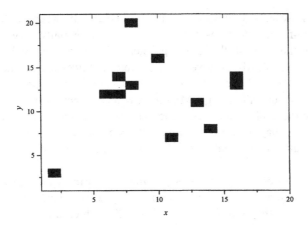

Figure 4.9. An approximate map. As $f(x,y)$ is likely to be such that enumeration at all possible points (x,y) is unfeasible, Figure 4.8 is unlikely to be realisable. Hence the requirement is to find a map giving the approximate locations of local optima.

In nature the exploration of separate fractions of the search space (or niches) by subsets of the population (or species) occurs readily. In applying this to the GA, the two most important concepts turn out to be *fitness sharing* and restrictions on who can mate with whom. That mating has to be restricted to closely related strings is not altogether surprising, it is after all one of the definitions of a species. That sharing the fitness between strings is important is more surprising.

SHARING

The importance of sharing can be visualised by consideration of the two-armed bandit, or fruit machine illustration used in reference [GO89, p186-189]. It is assumed that a team of players is playing a single machine. The question now arises, how many players should be allocated to each arm? If both arms pay out similar sums, at similar intervals, then the problem is trivial. However, if one (unknown) arm pays out more than the other (but with identical frequency), should all players play this arm, or should some play one arm and some the other? If they all play the arm which pays the larger prizes the winnings will be greater, but the money will be divided between more players. Whether the team should divide itself and allocate differing numbers of players to each arm or not depends on how any winnings are divided, or shared, between the team. If it is simply a free-for-all, or the winnings are carved up equally between all team members at the end of the evening, then all the players will migrate to whichever arm is slowly discovered to be better. However if a rule exists that monies should only be divided between players of the same arm, then a different behaviour will emerge. Migration will still occur to the better arm but after a while those playing the less good arm will learn that, although they win less as a group, they gain more individually because there are less members to share the winnings between. In short, they will have discovered that it makes sense to exploit a niche.

The ratio of team members playing each arm is easy to calculate. If there are 25 members in the team and arm A pays £50 and arm B £40, then if all players play on A they will receive £50/25 = £2 each on a pay-out. If they play B they receive £40/25 = £1.60 each; whereas, if a <u>single</u> member decided to play B they would receive £40/1 = £40. This is a very good incentive to desert A. A balance will occur once the <u>individual</u> pay-out from each arm is identical. This will occur when

$$\frac{Payout_A}{m_A} = \frac{Payout_B}{m_B}$$

where m_j is the number of players on arm j. In this illustration this will be when

$$\frac{£50}{m_A} = \frac{£40}{m_B}.$$

As

$$m_A + m_B = 25$$

substitution gives

$$\frac{50}{25 - m_B} = \frac{40}{m_B}$$

or

$$m_B = \frac{40 \times 25}{50 + 40} = 11.1$$

and therefore,

$$m_A = 25 - 11.1 = 13.9$$

With this split between the two arms, the pay-out will be £3.60 each, greater than the £2.00 each would receive if they all played A.

So, the introduction of sharing allows different niches to be exploited to the benefit of all. Different sub-populations, or species, exploit these niches, with the number of individuals within a niche being proportional in same way to the fitness of the niche. If this result could be generalised to other problems a powerful optimisation approach will have been developed. Not only would the filtering and mapping of the local optima described earlier have been achieved, but the number of individuals exploring any single peak would be proportional to the height of the peak. This seems a sensible approach as it will ensure the global optimum will still receive more attention than a minor local optimum. In order to develop these ideas further, the meaning of the term *species* in this context will have to be clarified.

SPECIES

In the natural world mating rarely occurs across species boundaries. So far, the genetic algorithms considered have had no restrictions on who can mate with whom. That there might be advantages, for some problems, in introducing restrictions can be understood by considering how a single point crossover might proceed in the search space shown in Figure 4.10.

If this function is mapped to a binary 4-bit string then $-1 = 0000$ and $1 = 1111$. These values both have maximum fitness ($f(x) = 1$), and thus one would expect any selection system to frequently pick them to mate. Unfortunately, such crossovers will often generate very sub-optimal strings. For example the crossover,

> *00/00*
> *11/11*

produces

> *0011*

and

> *1100*

as children. Neither of these is anything near optimal. In general, in a complex landscape, matings between distant, well performing, individuals will frequently produce poor offspring. There might therefore be some benefit in ensuring that only like mate with like.

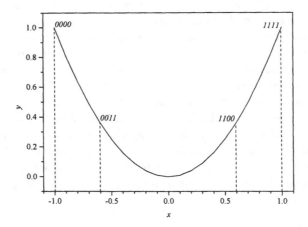

Figure 4.10. The production of sub-optimal solutions from the crossover of highly performing strings.

Goldberg and Richardson [GO87] introduced the idea of using a phenotypic-based sharing mechanism to re-allocate fitness within the population. The method makes use of a sharing function. This function is used to reduce the fitness of individuals who have a large number of close relations (phenotypically) within the population. This limits the uncontrolled growth of any particular species within the population; it also encourages the exploration of the whole of the search space and allows small populations of individuals to reside in any local optima discovered. The value of the sharing function s_i for an individual i depends on a sum of sharing values ξ_{ij}, between the individual and all other population members:

$$s_i = \sum_{j=1}^{N} \xi_{ij} \; .$$

The value of ξ_{ij} itself depends on the phenotypic distance between the two individuals i and j. Several possibilities have been suggested and Figure 4.11 illustrates one possibility.

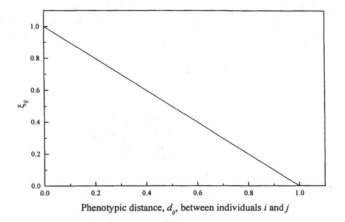

Phenotypic distance, d_{ij}, between individuals i and j

Figure 4.11. One possible sharing function. The distance d_{ij} is given by the absolute difference between the two phenotypes in this one parameter problem—the maximum difference being unity. (After [GO89 and GO89a]).

The method is implemented by temporarily reducing the fitness of each individual temporarily to f^{share}, given by:

$$f_i^{share} = \frac{f_i}{s_i} \; .$$

A successful application of the technique has been the work of Walters, Savic and Halhal [HA97] who have used sharing with multi-objective problems within the water industry (see §6.8).

As mentioned earlier, if lethals are to be avoided then some form of restrictions on mating may be required [HO71,DE89]. Alternatively, in a similar manner to sharing, Eshelman and Schaffer [ES91,ES91a] bar mating between similar individuals in an attempt to encourage diversity. Yet another possibility is to only allow fit individuals, in particular the elite member, to be picked once by the selection mechanism in order to slow convergence. Mahfoud, in [MA95], compares several niching methods.

4.3 CONSTRAINTS

Constraints can been visualised as giving rise to regions of the search space where no natural fitness can be assign. Such regions produce "holes" in the fitness landscapes (Figure 4.12). The question then arises of how to steer the GA around such holes. Lightly constrained problems pose few difficulties for GAs: the chromosome is decoded to a solution vector, which is in turn used within the problem to evaluate the objective function and assign a fitness. If any constraint is violated the fitness is simply set to zero.

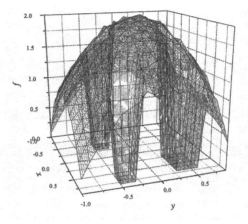

Figure 4.12. A fitness landscape with three large holes caused by the presence of constraints within a two-dimensional problem.

Although attractive, this approach is unlikely to be successful for more highly constrained problems (see [MI91] for some ideas). In many such problems the majority of possible solution vectors will prove to be infeasible. Even when this is not so, infeasible solutions may contain much useful information within their chromosomes. An alternative approach is to apply a *penalty function* [RI89] to any solution that violates one or more constraints. This function simply reduces the fitness of the individual, with the amount of reduction being a function of the violation.

The form of the penalty function must be chosen with care to maintain the correct balance between exploitation and exploration.

A further approach is the use of problem dependent crossover and mutation operators which do not allow the formation of infeasible solutions, for example the crossover operator introduced at the beginning of this chapter when discussing combinatorial optimisation.

Another approach is illustrated in the work of Walters, Savic and Halhal [HA97] where a messy GA [GO89a,GO91a,GO93] is used to build increasingly complex solutions from simple solutions that are known to be feasible (see §6.8).

In reference [PE97] Pearce uses a technique based on fuzzy logic to resolve constraints within a GA environment and discusses constraint resolution in general. Powell and Skolnick, in [PO93], and Smith and Tate, in [SM93], make general comments on non-linear constraints and penalty functions respectively. Reference [MI95] discusses the strengths and weaknesses of several approaches.

4.4 MULTICRITERIA OPTIMISATION

The optimisation problems considered so far have been expressed in a form where, although many parameters might be being optimised in parallel, the fitness of any particular solution can be specified by a single number. Not all problems share this attribute. In some problems the success of a particular solution can be estimated in more than one way. If these estimations cannot be combined, then a single measure of the fitness will be unavailable.

An example might be an attempt to minimise the cost of running a chemical plant: some of the possible operational strategies for the plant which reduce the financial cost of production might have the side-effect of increasing the likelihood of accidents. Clearly these solutions need to be avoided, whilst at the same time minimising the production cost in so far as practicable. Most importantly, solutions which are simultaneously better at minimising costs and reducing accidents need to be identified. The concept of *Pareto optimality* [GO89] can be used to identify such solutions. Figure 4.13 shows six possible strategies for operation of the fictitious plant.

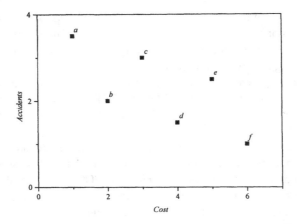

Figure 4.13. Six strategies for the operation of a chemical plant.

Solution *a* is optimal in terms of cost; *f* in terms of number of accidents. Solutions *c* and *e* are termed *dominated* because other solutions can be identified that simultaneously offer both fewer accidents and reduced cost, these are the *nondominated* solutions.

If estimations are made for a large number of operational strategies then the scatter plot of the outer points might take on the form of Figure 4.14.

The Pareto optimal set is then the set of all nondominated solutions on the inner edge of the scatter. Having identified this set (or the equation of the curve, or front, joining them) it is up to the management and workforce of the plant to settle on a particular strategy, drawn from this set.

Pareto optimality can be used in at least two ways to drive a rank-based selection mechanism within the GA. Nondominated sorting [SR94] identifies the Pareto optimal set and assign all members the rank of 1. These individuals are then removed from the ranking process. The next optimal front is identified and its members given the rank of 2. This process is repeated until all individuals have been ranked. BASIC code to carry out this procedure is given in Figure 4.15 and the approach is demonstrated in §6.8. An alternative approach is the Pareto ranking scheme of reference [FON93] where rank is proportional to the number of individuals that dominate the individual in question (Figure 4.16).

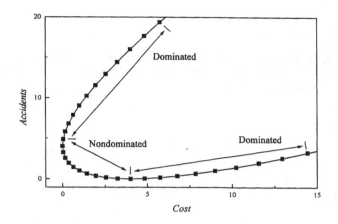

Figure 4.14. Surface formed from all possible strategies for the chemical plant's operation (only the outer points are shown). The Pareto optimal set (or front) is formed from the nondominated solutions.

```
FOR iRank = 1 TO N 'Loop over all possible ranks.
   FOR i = 1 TO N 'Pick an individual.
      IF Rank(i) = 0 THEN 'Only process unranked individuals.
         FOR j = 1 TO N 'Loop over all other individuals.
            IF i <> j THEN 'Check for domination.
               IF F1(i) < F1(j) AND F2(i) < F2(j) THEN EXIT FOR
            END IF
         NEXT j
         IF j = N + 1 THEN Rank(i) = iRank 'A nondominated solution has
      END IF                                              'been identified
   NEXT i
'Now remove the current nondominated front by setting the fitness of individuals on the front
   FOR i = 1 TO N                                                           'to zero.
      IF Rank(i) = iRank THEN
         F1(i) = 0
         F2(i) = 0
      END IF
   NEXT i
NEXT iRank
```
Figure 4.15. Nondominated sorting (continues over).

```
FOR i = 1 TO N 'Re-assign fitness based on rank.
  F(i) = 1 / Rank(i)
NEXT i
```

Figure 4.15 (continued). BASIC code to carry out nondominated sorting of a population of size N. The problem contains two measures of fitness, F1 and F2, which are reduced to a single measure F by letting F =1/Rank.

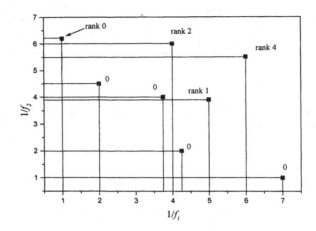

Figure 4.16. Pareto ranking for a problem of two criteria giving rise to two fitness functions f_1 and f_2.

Both techniques require the use of fitness sharing to ensure that the population covers a large enough fraction of the search space [HA97]. (See [GR97] for some recent ideas about this).

4.5 HYBRID ALGORITHMS

Genetic algorithms are not very good at finding optimal solutions! However they are good at navigating around large complex search spaces tracking down near-optimal solutions. Given enough time a GA will usually converge on the optimum, but in practice this is not likely to be a rapid process. There are many other, more efficient, traditional algorithms for climbing the last few steps to the global optimum. This implies that a very powerful optimisation technique might be to use a GA to locate the hills and a traditional technique to climb them. The final algorithm will depend on the problem at hand and the resources available.

The simplest approach to this hybridisation is to use the real-valued solution vector, represented by the fittest individual in the final generation, as the starting point of a traditional search. The traditional search algorithm could be either a commercial compiled one, such as a NAG routine, or taken from a text on numerical methods.

Another, approach is to stay with the string representation used by the GA and attempt to mutate the bits in a directly constructive way. One way to achieve this is illustrated in the estimation of the ground-state of a spin-glass presented in §6.4. In this example, local hills are climbed by visiting each bit within the string in turn, mutating its value and re-evaluating the fitness of the population member. The mutation is kept if the fitness has improved. Another, very simple, possibility is to hill-climb by adding (or subtracting) *1* to the binary representation of the first unknown parameter in the elite string (e.g. *1101 + 1 = 1110*), re-evaluating the fitness, and keeping the addition (subtraction) if it has proved beneficial. This addition (subtraction) is repeated until adjusting the first unknown parameter shows no benefits. The other parameters are then treated in the same way.

Working with the GA strings themselves has the advantage that such techniques can be applied at any moment during a genetic algorithm's run. Moving to a real encoding can make it difficult to return to a binary string represented GA, because some parameters may have taken values that can not be represented directly by such a string (see Chapter 2). However, such real-valued methods are typically highly efficient. One way around this problem is not to use a binary string representation within the GA (as discussed in §4.7).

If the search space is believed to be complex at many scales, abandoning the GA in favour of another method too soon can lead to an erroneous solution. The liquid crystal problem studied in §6.5 contains just such a space. In this work, using the final solution vector as the starting point for a more constrained GA-based search was found to be effective.

Other methods of improving performance and convergence speed make use of heuristics. One such example is the use of inter-city distances within a TSP (i.e. making it no longer blind). Grefenstette et. al. used this information to produce an improved uniform-type crossover operator. Rather than building child strings by taking alternating cites from each parent, the child inherits the city which is geographically closest to the current city [GR85].

Alternatively, the fitness evaluations—which are typically the most time consuming element of the algorithm—can initially be done in an approximate manner (see §6.5). For example, in problems which use least-squares minimisation of experimental data, this can be achieved by initially only presenting the GA with a random, regular or other subset of the data and running a fixed number of generations. The initially subset is then enlarged to include more of the data and further generations processed [MIK97a]. This process is repeated until all the data is being considered.

4.6 ALTERNATIVE SELECTION METHODS

The selection pressure within the GA is central to its performance, and the appropriate level is highly problem dependant. If the population is pushed too hard, rapid progression will be followed by near stagnation with little progression in f_{max}. This is unsurprising. With a high selection pressure the population will become dominated by one, or at most a few, super-individuals. With very little genetic diversity remaining in the population, new areas of the problem-space become unreachable—except via highly unlikely combinations of mutations. Another way of visualising the effect of this pressure is by considering how directed the mechanism is toward a sub-set of the population (typically the best). Highly directed mechanisms will result in a path-orientated search, less directed mechanisms will result in a volume-orientated search.

The selection pressure can be characterised by the take-over time, Γ [GO91]. In essence, this is the number of generations taken for the best individual in the initial generation to completely dominate the population. (Mutation and crossover are switched off). The value of Γ depends not only on the selection mechanism, but for some mechanisms, on the function being optimised. If fitness proportional selection is used, then for:

$$f(x) = x^a, \ \Gamma \approx \frac{1}{a}(N \ln N - 1)$$

and for

$$f(x) = \exp(ax), \ \Gamma \approx \frac{1}{a} N \ln N$$

[GO91,BA96,p168], i.e. of the general order $N \ln N$.

Other selection mechanisms are common (see [GO91]) and in essence they all try to encourage the GA to walk a fine tight-rope between exploitation and exploration, whilst minimising sampling errors. Such mechanisms usually make the assumption that if a individual has a lower fitness, it is less likely to be selected. This need not be so, as Kuo and Hwang point out in [KU93].

STOCHASTIC SAMPLING ERRORS

Fitness-proportional selection is a stochastic method. On average, the number of trials, τ_i, (in the next generation) an individual, i, is given will be such that if

$$f_i = 2f_{ave}$$

then

$$\tau_i = 2\tau^{ave}$$

where τ^{ave} is the number of trials an individual of average fitness would achieve (typically 1). This number will not always be achieved. GAs make use of numerous calls to random number generators, and given enough calls, some surprising patterns can emerge. Table 4.2 and Figure 4.17 show the result of using the roulette wheel algorithm in LGADOS for $N = 20$ and $G = 200,000$. The results are for the elite member which, because of the problem used, has $f(g) = 5f_{ave}(g)$ for all g, and should thus have on average five trials in the next generation. Although five is indeed the most common number of trials allocated, many other possibilities also occur. On 621 occasions no trials were allocated to this the fittest member. This implies that, unless elitism is in place, the best solution would be discarded. Conversely f_{max} will occasionally almost flood the population. Such over or under selection can greatly impair the performance of the algorithm, and a number of alternative selection methods have been designed to minimise the problem.

Number of trials	Expected frequency	Actual frequency	Number of trials	Expected frequency	Actual frequency
0	0	621	9	0	5286
1	0	4166	10	0	2023
2	0	13318	11	0	587
3	0	26922	12	0	160
4	0	38036	13	0	17
5	200000	40334	14	0	2
6	0	33714	15	0	2
7	0	22599	16	0	0
8	0	12213			

Table 4.2. Stochastic sampling errors as shown by the difference between the expected number of trials and the actual number granted by fitness-proportional selection ($f = x^2$, $N = 20$, $G = 200000$, $\acute{\varepsilon} = 0$, $\mathit{R} = 1$).

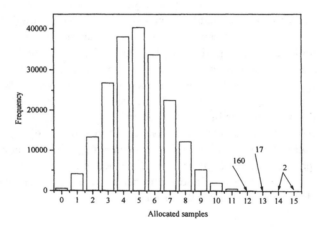

Figure 4.17. The distribution of samples of the elite member found; the expected number should always be five in this problem.

STOCHASTIC SAMPLING

Stochastic sampling with replacement is another name for roulette wheel selection. Each time an individual is selected it is also returned to the pool—allowing for multiple selections of the same individual. As discussed above, this replacement can result in:

$$\tau_i(g) >> \tau_i^{exp}(g) \ ,$$

(the expected number of trials for individual *i*). *Stochastic sampling without replacement* forces the maximum number of trials an individual can receive to equal unity (by not returning selected individuals to the pool). This is a major brake on the selection mechanism. However, it does still allow τ_{best}, the number of trials allocated to the fittest individual to equal zero occasionally.

Remainder stochastic sampling (with and without replacement) estimates $\tau_i^{exp}(g)$ then sets:

$$\tau_i(g) = INT\big(\tau_i^{exp}(g)\big) \ ; \ i=1...N \ ,$$

where *INT*(-) returns the integer part of (-).

In general this will leave some slots in the new population unfilled. Stochastic sampling with replacement is then used to fill the remaining positions by using the fractional parts.

$$\tau_i^{exp}(g) - INT\big(\tau_i^{exp}(g)\big) \ ; \ i=1...N \ ,$$

to assign roulette wheel slots.

The method can also be used without replacement by simply using a random number R^+, between 0 and 1. An individual is selected if:

$$R^+ \geq \tau_i^{exp}(g) - INT\big(\tau_i^{exp}(g)\big) \ .$$

Stochastic universal sampling [BA87] also uses a roulette wheel but with N equal spaced markers. The wheel is spun only once and all individuals which fall adjacent to a marker are selected.

RANKING METHODS

If the position of an individual within a list ordered by fitness is used to calculate τ_i then problems of super-fit individuals are avoided. The position of the individual within the list is all that matters, not how much fitter than the population average it may be. This greatly suppresses problems of premature convergence, whilst still providing a suitable level of selection pressure in later generations.

In its simplest form, the population is ranked and the best 50% of individuals selected to go forward to the next generation. Crossover is then performed on all selected individuals by choosing random pairs without replacement.

More subtle methods have been presented [BA85]. One possibility is to fix τ_{best}^{exp} by hand and then apply a linear curve through τ_{best}^{exp} that encloses an area equal to N. The height of the curve then supplies τ_i^{exp} for each individual (Figure 4.18). One problem with this approach is the need to select the value of τ_{best}^{exp}, which will (of course) be problem dependent. Other methods use a non-linear curve [MI94].

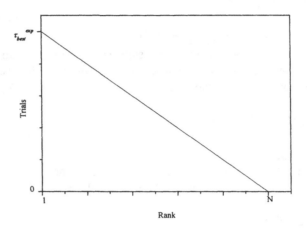

Figure 4.18. Linear ranking.

The take-over time for rank-based selection depends on the details of how it is applied, in particular on the value of τ_{best}^{exp}, but Γ is of the order lnN, i.e. much lower than with fitness-proportional selection [GO91,BA96,p171].

TOURNAMENT SELECTION

Tournament selection [GO91,BL95] is both effective and computationally efficient. Pairs of individuals are selected at random and a random number, R^+ (in the range 0-1) generated. If $R^+ > r$, $0.5 < r \leq 1$, then the fitter of the two individuals goes forward, if not, the less fit. The value of r also requires setting by hand.

In other implementations q individuals are initially selected with the single best going through to the next generation. Such an approach has:

$$\Gamma \approx \frac{1}{\ln q}\left(\ln N + \ln(\ln N)\right) \ .$$

This implies that the take-over time will rapidly decrease as q moves away from 2 (i.e. binary tournaments).

SIGMA SCALING
Linear fitness scaling (Chapter 3) can be extended by making use of the population fitness standard deviation f^{σ} [MI96], with the expected number of trials given by:

$$\tau_i^{exp} = 1 + \frac{f_i(g) - f_{ave}(g)}{2f^{\sigma}} \ .$$

STEADY-STATE ALGORITHMS
LGA is a *generational* algorithm in that at each generation, a new population is formed (although some will be direct copies of their parents not disrupted by crossover or mutation). Steady-state algorithms [SY89,SY91,WH89,DE93a] replace only a few of the least fit individuals each generation by crossover and mutation, and thus require few fitness evaluations between generations. The fraction of individuals replaced is called the *generation gap* [DE75]. Such algorithms have proved highly effective on problems where identical genotypes always return the same value of f for all estimations (this will not necessarily be so with noisy, time varying data) [DA91].

4.7 ALTERNATIVE CROSSOVER METHODS

Single point crossover has been criticised for several reasons [CA89, ES89, SC89a]. Although it can recombine short, low-order, schemata in an advantageous manner, it frequently cannot process all schemata in the same way. For example, given:

010####1 and
###00###,

single point crossover cannot form

01000##1.

Such *positional bias* [ES89] implies that schemata with long defining lengths suffer biased disruption.

Crossover of strings that are identical to one side of the cut-point will have no effect, as the children will be identical to the parents. The reduced surrogate operator [BO87] constrains crossover so as to produce new individuals whenever possible. This is achieved by limiting cut-points to points where bit values differ.

TWO-POINT CROSSOVER

In order to reduce positional bias, many researchers use two-point crossover where two cut points are selected at random and the portion of chromosome between the two cuts swapped. For example:

00/0100/111 and
11/1011/000 give

001011111 and
110100000.

Just as with single point crossover, there are many schemata which two-point crossover cannot process, and various numbers of cut points have been tried [EC89].

UNIFORM CROSSOVER

Uniform Crossover [SY89] takes the idea of multi-point crossover to its limit and forces an exchange of bits at every locus. This can be highly disruptive, especially in early generations. *Parameterised crossover* [SP91] moderates this disruption by applying a probability (typically in the range 0.5-0.8) to the exchange of bits between the strings. (For a discussion of multi-point crossover see [SP91b]).

4.8 CONSIDERATIONS OF SPEED

In most practical applications, the time taken to estimate the objective functions, Ω, will be greatly in excess of the time taken to carry out any genetic operations. Therefore there is little need to worry about trying to time-optimise these operations. One way to ensure minimum run times is to try and speed the estimation of $\Omega_i(g)$. Two possibilities are to interpolate from pre-existing values, or to use an approximate value of $\Omega_i(g)$ for some generations but not others. This could be by using the approximation for $g < g'$ then reverting to the true estimation for $g \geq g'$. This approach is used in §6.5, when only a sub-set of the experimental data is used in early generations.

An obvious but important possibility is to ensure that $\Omega_i(g)$ is not re-estimated once it has been found for a particular chromosome. This will require maintaining a (possibly ordered) list of all values of Ω calculated during the run. Clearly, this list has the potential to become extensive and it might be of value to only store it for $g - 1$ to $g - k$, where the historic period k used will depend on the relative time overhead of examining the list and estimating another value of Ω. If new estimates are placed at the bottom of the list, it will probably prove worthwhile to search the list in reverse order to maximise the gain (see §6.6).

A further possibility is to use a network of computers to carry out separate estimates of $\Omega(g)$ on separate machines.

In addition to all of these considerations, it is necessary to ensure the use of the minimum values of l_j $(j = 1...M)$ and to keep the range of values each parameter can take as small as possible. Both the range and the string length can be functions of generation (see §6.5) but care must be used as this effectively removes areas of the search space during certain generations. Alternatively a messy-GA [GO89a,GO91a] can be used to build complex solutions from simple building blocks (for example, §6.8).

4.9 OTHER ENCODINGS

This text has concentrated on binary encoded GAs. Many authors have pointed out that GAs will probably be at their most effective when the encoding is as close as possible to the problem space. For many problems in science and engineering this implies the use of numbers of base-10 form. Unfortunately, using a real-valued encoding poses a large number of questions; in particular, what to use as crossover and mutation operators. Several possibilities have been promoted and a detailed discussion is to be found in references [MI94], [ES93], [JA91] and [WR91]. Reeves in [RE93] makes comments on the implications for population sizes of using non-binary alphabets.

One possibility for crossover (between individuals i and k) is to use [MU93,ZA97, p14-16]:

$$r_i(g+1) = r_j(g)[r_k(g) - r_i(g)]R \; ; k \neq j$$

where R is a random scaling factor (typically in the range -0.25 to 1.25).
Mutation can be included in several ways, for example:

$$r_i(g+1) = r_i(g) + R^{\pm}(g)r_i(g) \; ; R(g=0) = R^{\pm}, R(g) \rightarrow 0 \text{ as } g \rightarrow G .$$

Things need not however be made too complex and a binary representation will often work extremely well. As Goldberg has pointed out [GO89,p80], GAs are typically robust with respect to the encoding used. He gives two simple coding rules:

1. **The Principle of Meaningful Building Blocks**: the user should select a coding so that short, low-order schemata are relevant to the underlying problem and relatively unrelated to schemata over other fixed positions.

2. **The Principle of Minimal Alphabets**: the user should select the smallest alphabet that permits a natural expression of the problem.

It is relatively easy to get some idea of why the use of a binary encoding is a reasonable choice for many problems. Consider some of the strings that might occur when solving $f(x) = x^2$, $0 \leq x \leq 15$ with either a 4-bit binary representation or a one-to-one mapping of binary integers to the first 16 letters of the alphabet. Table 4.3 shows five possible values of x together with their respective binary and non-binary strings. As the list is descended, there is an obvious visual connection between the binary strings of fitter individuals made possible by their low cardinality (number of distinct characters): they all have 1's toward the left-hand side of the string. In the non-binary case no such similarities can exist. As these similarities are at the heart of the search method, their number should be maximised.

Another way to emphasise the difference is to count the number of schemata available in the two representations. For cardinality k, this will be $(k + 1)^L$. As the same accuracy is required from both encodings, $L = 4$ in the binary case and 1 in the non-binary case. Therefore, the binary representation contains $(2 + 1)^4 = 81$ schemata and the non-binary $(16 + 1)^1 = 17$ schemata (a to n plus #), a clear advantage to the binary representation.

x	f	C(binary)	C(non-binary)
1	1	*0001*	*a*
2	4	*0010*	*b*
3	9	*0011*	*c*
12	144	*1100*	*l*
14	196	*1110*	*m*

Table 4.3. Comparison of binary and non-binary representations.

LOGARITHMIC REPRESENTATION

For many scientific and engineering problems, it is only the relative precision of the unknowns which is important [ZA97]. In such cases the encoding of the logarithm of the unknown may prove more suitable. This will allow a very large space to be searched without the need for excessive string lengths, but still maintain reasonable accuracy when needed. For example, if an unknown, r, may take any value between 1 and 1,000 and the answer is required to 1 part in 100, then a simple linear mapping between the problem space and a binary representation requires adjacent points to vary by only $1/100 = 0.01$. This will then allow values of r around 1 to be distinguished to the required accuracy of 1%. However, the precision around $r = 1,000$ will still be 0.01, which is 1 part in 100,000 or 0.001%. This implies that most of the string is redundant.

By using a logarithmic mapping the required accuracy can be maintained throughout the space with a much shorter string, thereby enhancing performance.

GRAY ENCODING

It was suggested above that a GA might be at its most successful when the encoding used is as close to the problem space as possible. Thus, a "small" change in the phenotype should indicate an equally "small" change in the genotype. For a binary encoding this is not so. Given a genotype with $l = 6$ and a phenotype, r with $0 \leq r \leq 63$ then

$$011111 = 31$$

will have to undergo changes to all six bits to increase the value of the phenotype by one:

$$100000 = 32.$$

Gray binary encoding alleviates this by ensuring that any pair of adjacent points in the problem space differ by only a single bit in the representation space. (Note, a small change in the genotype can still lead to a very large change in the phenotype). For many problems this adjacency property is known to improve performance and it would seem sensible to adopt the use of a Gray encoding for most problems with integer, real or complex-valued unknowns.

Table 4.4 list binary and Gray equivalents for $l = 4$. Figure 4.19 presents BASIC code for converting Gray to standard binary; reference [ZA97, p96] gives pseudo-code for carrying out this transformation.

Binary	Gray	Binary	Gray
0000	0000	1000	1100
0001	0001	1001	1101
0010	0011	1010	1111
0011	0010	1011	1110
0100	0110	1100	1010
0101	0111	1101	1011
0110	0101	1110	1001
0111	0100	1111	1000

Table 4.4. Comparison of binary and Gray encoding.

```
Bin(1) = Gray(1)

FOR i = 2 TO L
  IF Bin(i - 1) = Gray(i) THEN
    Bin(i) = 0
  ELSE
    Bin(i) = 1
  END IF
NEXT i
```

Figure 4.19. BASIC code to convert Gray to standard binary. The Gray and binary strings, Gray and Bin (each of length L) are assumed to be held in arrays where the first element is the most significant.

4.10 META GAs

Much has been made in this text of the need to choose the internal GA settings (P_c, P_m, N, etc) with care, and of the fact that their optimum settings are likely to be highly problem-dependent (as will be the exact form of the algorithm). This leads to the very natural question, why not have these settings selected by a separate GA? Furthermore, as $f_{best} \rightarrow f^*$ the form of the search space is likely to change, implying that the ideal internal settings might not be constant in time. This leads naturally to the possibility of using a GA to optimise the internal settings during the run itself.

If only a single value of each internal parameter is required, and the GA is going to be used repeatedly with near-identical problems, then the approach is relatively easy to implement by using a meta-level GA [GR86,BR91] to control and evolve settings within a population of GAs.

However in difficult problems, or where the GA will only be used a few times, the additional computing required probably makes the approach unrealistic. The ability to allow the settings to adapt in real-time [DA89,DA91], might however allow the design of extremely efficient algorithms, and is a particularly interesting avenue of research.

4.11 MUTATION

So far mutation has come in a single guise: the infrequent random flipping of bits within the chromosome. Traditionally, mutation has always been seen as very much a secondary mechanism in comparison to crossover. However there would now appear to be a growing feeling that it may have a more central role to play [MU92a,HI95,JO95].

The mutation operator as described in Chapter 1 is rather a blunt instrument, the main role of which would appear to be ensuring that the population maintains both possible bit values (*0* and *1*) at all loci. If mutation is to be applied in a more directed manner, i.e. as part of the search process itself, then it would seem sensible to make the operator more discriminating. For example, with a binary representation (standard or Gray) the magnitude of the disruption caused by mutation depends upon where in the chromosome mutation occurs. Given a single parameter problem with $L = 10$, $r_{min} = 0$ and $r_{max} = 1023$, then a mutation at one end of the chromosome changes r by ± 1, whereas at the other end r would change by ± 512.

This analysis indicates that near the end of the run, when hopefully the majority of the population is in the vicinity of the global optimum, there might be advantages in confining mutation to lower order bits. Conversely, during earlier generations mutation of the higher order bits will help in the full exploration of the space, whereas mutation of less significant bits will add little to this exploration. (See the exercises at the end of this chapter for one possibility).

Other possibilities are to bias mutation towards less fit individuals to increase exploration without degrading the performance of fitter individuals [DA89], or to make P_m a function of g with the probability decreasing with generation [FO89].

Alternatively, mutation can be used in a hill-climbing sense to close in on the final solution after $g = G$. This is achieved by using mutation to slowly increase (or decrease) estimates of each unknown within the elite member, keeping the change only if it has proved constructive (i.e. increased f_{max}).

See [BA93] and [TA93] for a discussion on setting optimal mutation rates.

4.12 PARALLEL GENETIC ALGORITHMS

Apart from the implicit parallelism provided by schemata processing, the population-based approach of GAs makes them ideal for implementation on parallel, or networked, machines. Although, for many, the reason for using a parallel implementation will be a simple increase in speed due to the increase in computer power, others use pseudo-parallel approaches that are believed to improve performance even when used on sequential machines.

Global Parallel GAs [GO89,HU91,DO91] treat the population as a single unit and assign different individuals to different processors. In its simplest form, the approach uses one machine (or processor) to control selection and the genetic operators, and a series of other machines (or processors) to carry out objective function evaluation. If the objective function evaluation takes a considerable time, a substantial speed-up can be achieved given a little additional code and a roomful of personal computers.

Migration or *Island* GAs [TA87,TA89] attempt to mimic the geographical separation of subpopulations witnessed in the natural world. This is achieved by allowing separate subpopulations or *demes* of chromosomes to evolve using selection and crossover, but then allowing occasional migration of individuals between subpopulations.

Diffusion, neighbourhood, cellular, or *fine-grained* GAs remove the subpopulation barriers used in the migration methodology and replace them with the concept of geographical *distance*. Individuals are then only allowed to breed with close neighbours [RO87,MA89,SP91a,DA91a,MA93].

Parallel GAs are introduced in greater detail in reference [ZA97, p20-30].

4.13 SUMMARY

In this chapter several advanced operators have been introduced, including ones to tackle selection and crossover in combinatorial and multicritia optimisation. The latter leads naturally to the introduction of the concept of maintaining sub-populations of differing species within the algorithm. One possibility for dealing with constraints, the penalty function, has been suggested—although in practice its use is far from straightforward.

Alternative representations have been introduced and the recommendation that if a binary-type representation is used, then a Gray encoding is adopted. The use of a logarithmic representation has also been promoted.

Hybrid algorithms, which combine a GA with a more traditional algorithm, have been hinted at as a highly powerful combination for solving practical problems.

Alternative selection, crossover and mutation mechanisms have been discussed, in part to alleviate stochastic sampling errors, but also as a way of ensuring the correct balance between exploration and exploitation. Meta GAs, where the algorithm itself adapts during a run, are one way this balance might be naturally found.

Finally, parallel algorithms have been mentioned as an interesting avenue of research.

4.14 EXERCISES

1. Introduce PMX into LGADOS.BAS, or your own GA, and use it to help solve the simple TSP detailed in Table 4.1.

2. By including a subroutine to apply a simple penalty function, solve a lightly constrained multi-dimensional problem of your choice. Experiment with the strength of the penalty function to see the effect on the efficiency of the algorithm. Increase the level and number of constraints until most of the search space is unfeasible. (This should indicate that penalty functions are possibly inappropriate for such problems.)

3. Introduce tournament selection, linear rank selection, two-point crossover and uniform crossover into LGADOS.BAS, or your own GA, and study their effect on test function performance.

4. Design and implement a new mutation operator, P'_m, which gives a reduced probability of mutation of more significant bits during later generations: $P'_m = P_m \Xi(j,g)$; $1 \leq j \leq l_k$, $1 \leq k \leq M$. The function Ξ should be linear in j. Study the effect of differing functional forms on test function performance.

5. Convert LGADOS.BAS, or your own GA, to a Gray encoding. (This is easier than it might seem and only involves changing how the binary strings are converted to integers.) Compare the performance of binary and Gray encoding on various test functions and various mutation rates.

6. Allow for a logarithmic representation within LGADOS.BAS, or your own GA. Use the new code to solve MAX[$f = x$]; $0 \leq x \leq 1 \times 10^6$ to 1% across all x and compare the performance to a linear representation.

7. Adapt LGADOS.BAS, or your own GA, to stop the evaluation of individuals that existed in the previous generation. Plot graphs of f_{max} against number of objective function evaluations for some of the test functions of Chapter 3. What effect does this have on the efficiency of the algorithm for various settings of P_c and P_m? Extend the approach to stop the evaluation of individuals that have occurred in the last k-generations ($1 \leq k \leq g$). Use $N \geq 100$ in your experiments.

CHAPTER 5

WRITING A GENETIC ALGORITHM

Implementing a genetic algorithm on a computer is surprisingly easy. Much of this simplicity arises from the lack of sophisticated mathematics within the algorithm: there are few operations that would be beyond even the most rusty or inexperienced of individuals.

For most applications, the programming language chosen for the GA itself will be of little relevance because the majority of computational time will be spent estimating the objective function. Thus it is far more important to ensure that that part of the program is optimised in structure and language. As most programming environments allow mixed language programs, the GA and the objective function routines need not even be in the same language (or even running on the same machine).

Although there are several GA packages available, either commercially or for free, I strongly believe in the value of trying to code one's own simple algorithm, at least in the first instance. For experienced programmers this will take less than a day.

The implementation described here is of LGADOS and is designed to run under QBASIC on a PC. BASIC has been chosen, in part, because of its ease of comprehension by the less experienced (and because it is included on most DOS-based systems). Sophisticated data structures have been avoided, and although this possibly makes the algorithm slightly less elegant, it should allow for easy translation into any other language (translations into PASCAL, FORTRAN and C are already provided on the disk). Those wishing to work in PASCAL might like to use code from reference [GO89]; those with a preference for C could use the code in reference [MI94] (although this is for a real-valued, rather than a binary, encoded GA).

This chapter starts by sketching the form of the program, explaining the data structure adopted, listing the main program and then examining each operation and its associated subroutine (procedure) in turn. Extracting the results and adapting the program to solve other problems is also discussed. Although a rough understanding of the program is advisable, it would be possible simply to make suitable adjustments, as detailed below, and run the code via QBASIC or a BASIC compiler (QuickBASIC would be ideal) and solve other, more complex, problems.

A complete listing of the program is included on the disk and in Appendix B.

A SKETCH OF THE PROGRAM

At the heart of the program are two non-overlapping populations of binary encoded strings. One is the current generation, the other a new temporary population in the process of being constructed from the current generation by selection, crossover and mutation. When the temporary population is complete it replaces the current generation and the generational counter is incremented by one. This process is sketched in Figure 5.1.

```
Generation = 1.
Create initial population 'Build a population of strings at random.
Find unknowns 'De-code the new population to integers then real numbers.
Find fitness 'Find the fitness of each member of the population.

FOR Generation = 2 TO maximum number of generations

   FOR NewIndividual = 1 TO PopulationSize STEP 2    'Loop over the
                                        population choosing pairs of mates
      Select a mate
      Select other mate
      Perform crossover    'Pass individuals to the temporary population after
                              performing crossover.
   NEXT NewIndividual

   Mutate   'Mutate the temporary population.
   Replace  'Replace the old population completely by the new one.
   Find unknowns   'De-code the new population to integers then real numbers.
   Find fitness    'Find the fitness of each member of the population.

NEXT Generation
```

Figure 5.1. A sketch (not real code) of the program.

The code fragment:

```
FOR k = 1 to 10
     ~ ~ ~ ~ ~
     ~ ~ ~ ~ ~
NEXT k
```

is BASIC's way of implementing a loop a fixed number of times (in this case 10). The option "STEP" can be used to increment the counter by more than unity each time. A single quote in BASIC indicates that any characters to the right are comments. Some of the lines of the program have been split in order to fit them on the page; such wrap-arounds are indicated by the symbol Ξ if they occur within code but not indicated if they occur within comments.

The program starts by building a population of random strings (Create initial population); these are then each converted to real-valued solution vectors (Find unknowns) which are tested on the problem at hand and assigned a fitness (Find fitness). This first generation then undergoes selection (Select a mate and Select other mate) and crossover (Perform crossover) a pair at a time and a new temporary population is constructed. The temporary population then undergoes mutation (Mutate) and replaces (Replace) the current generation. The new population members are then converted to real-valued solution vectors, tested on the problem and assigned a fitness. The program is finished when a set number of generations (maximum number of generations) have been processed.

Each of these operations are expanded upon below, but first the form of the arrays used to store the population must be described.

DATA STRUCTURES

Although the use of user-defined data types (or records) would make for a slightly more elegant program, these have been avoided in the interest of simplicity and to ensure ease of translation. The population is held in a series of arrays with each array representing one aspect of the population. Thus the binary strings are held in one array (called Strings) as rows of integers (each with the value 0 or 1). These are decoded as though they were true binary strings to an array called Integers and then to real-valued solution vectors in array Unknowns. The fitness of each individual is held in the single column array Fitness.

As an example of the layout of these arrays, consider a population consisting of four individuals and a problem ($f = x + y$), with each unknown being represented by a string of length three and the unknowns being in the range 0 to 14:

Strings:

$$\begin{bmatrix} 1 & 0 & 1 & 1 & 1 & 1 \\ 0 & 0 & 0 & 1 & 1 & 0 \\ 1 & 1 & 0 & 1 & 0 & 0 \\ 0 & 1 & 0 & 0 & 0 & 1 \end{bmatrix}$$

Integers:

$$\begin{bmatrix} 5 & 7 \\ 0 & 6 \\ 6 & 4 \\ 2 & 1 \end{bmatrix}$$

Unknowns:

$$\begin{bmatrix} 10 & 14 \\ 0 & 12 \\ 12 & 8 \\ 4 & 2 \end{bmatrix}$$

Fitness:

$$\begin{bmatrix} 24 \\ 12 \\ 20 \\ 6 \end{bmatrix}$$

Thus, using the notation of Chapter 2, for $i = 1$ (the first member), $C = 101111$, $z_1 = 5$, $z_2 = 7$, $r_1 = 10$, $r_2 = 14$ and $f = 24$.

The array NewStrings represents the new, temporary, population. EliteString, EliteIntegers, EliteUnknowns and EliteFitness hold a copy of the genotype, integer phenotype, real phenotype and fitness of the elite member.

An array, Range, holds the upper and lower bounds of each unknown. For the example above:

Range:

$$\begin{bmatrix} 0 & 0 \\ 14 & 14 \end{bmatrix}$$

The substring length l; G; P_c; P_m and c_m are all held as constants (i.e. they can not be altered during execution). All the above arrays and the constants are "shared", i.e. they are global variables which all parts of the program have automatic access to.

THE MAIN PROGRAM

The program consists of a "main" program and a series of subroutines "called" from the main program. The format of the call statement is:

```
CALL subroutine name (arg1, arg2, .......)
```

Such a statement transfers control to the subroutine and passes the arguments `arg1`, `arg2` etc to the subroutine. The names of the subroutines hopefully indicate their purpose.

The main program is listed in Figure 5.2. First the constants are defined, then the arrays are dimensioned and the bounds of each unknown stated. The random number generator is then "randomised" to ensure the program uses a different sequence of random numbers each time it is run and the files to hold the results opened. An initial population is then created and the fitness of the population members found. Subroutine `statistics` calculates the sum and mean of the fitness and finds which individual has the highest fitness. `PrintGeneration` does just that: it prints the results to the screen and to the two result files. `scaling` applies linear fitness scaling to each individual. Because this scaling is applied after `PrintGeneration` the results printed include the true fitness, not the scaled values.

After this first generation is complete, generations 2 to G are processed. Pairs of mates (or parents) are chosen by fitness proportional selection. A random number is then thrown, and if this is number less than or equal to P_c crossover is used to build the new strings, otherwise the strings are simply cloned. Mutation is then applied and `Replace` used to overwrite the old population with the new. The strings are then decoded to produce the unknowns, the fitness of each solution vector found and the results output. Subroutines are included to apply elitism and fitness scaling if required.

```
'------- SET ALL THE IMPORTANT FIXED PARAMETERS. -------

'These should be set by the user.
CONST PopulationSize = 20      'Must be even.
CONST NumberOfUnknowns = 2
CONST SubstringLength = 12     'All sub-strings have the same length.
CONST TotalStringLength = NumberOfUnknowns * SubstringLength
CONST MaxGeneration = 20 'G.
CONST CrossOverProbability = .6      'Pc >=0 and <=1.
CONST MutationProbability = 1 / TotalStringLength    'Pm, >=0 and <1.
CONST Elitism = "on"     '"on" or "off".
CONST ScalingConstant = 1.2    'A value of 0 implies no scaling.


'------DECLARE ALL SHARED (I.E. GLOBAL) VARIABLES----------

'The arrays that hold the individuals within the current population.
DIM SHARED Unknowns(PopulationSize, NumberOfUnknowns) AS SINGLE
DIM SHARED Integers(PopulationSize, NumberOfUnknowns) AS LONG
DIM SHARED Strings(PopulationSize, TotalStringLength) AS Ξ
                                                         INTEGER
DIM SHARED Fitness(PopulationSize) AS SINGLE


'The new population.
DIM SHARED NewStrings(PopulationSize, TotalStringLength) AS Ξ
                                                           INTEGER


'The array that defines the range of the unknowns.
DIM SHARED Range(2, NumberOfUnknowns) AS SINGLE


'The best individual in the past generation. Used if elitism is on.
DIM SHARED EliteString(TotalStringLength) AS INTEGER
DIM SHARED EliteIntegers(NumberOfUnknowns) AS LONG
DIM SHARED EliteFitness AS SINGLE
DIM SHARED EliteUnknowns(NumberOfUnknowns) AS SINGLE


CLS     'Clear the screen.


CALL DefineRange     'Define the range of each unknown. These should also be set by the
                      user.
```

Figure 5.2. The main program (continued over).

'Set the random number generator so it produces a different set of numbers
'each time LGADOS is run.
```
RANDOMIZE TIMER
```

```
CALL OpenFiles    'Open files used to store results.
```

'------ START OF THE GENETIC ALGORITHM ----------------------

'------ CREATE AN INITIAL POPULATION, GENERATION 1 ------

```
Generation = 1
```

```
CALL InitialPopulation    'Build a population of strings at random.
```

```
CALL FindFitness    'Find the fitness of each member of the population.
```

```
CALL Statistics(MeanFitness, SumFitness, FittestIndividual, Ξ
                Generation)    'Find the mean fitness and the fittest individual.
```

```
CALL PrintGeneration(Generation, MeanFitness, Ξ
                     FittestIndividual)    'Print generation to file.
```

```
CALL Scaling(ScalingConstant, FittestIndividual, SumFitness, Ξ
MeanFitness)    'If linear fitness scaling is "on" then scale population prior to selection.
```

'------ LOOP OVER ALL THE GENERATIONS ------

```
FOR Generation = 2 TO MaxGeneration
```

```
FOR NewIndividual = 1 TO PopulationSize STEP 2    'Loop over the
                                    population choosing pairs of mates.
```

```
    CALL Selection(Mate1, SumFitness, MeanFitness)    'Choose first
                                                       mate.
    CALL Selection(Mate2, SumFitness, MeanFitness)    'Choose second
                                                       mate.
```

'Pass individuals to the temporary population either with or without performing
crossover.
```
IF RND <= CrossOverProbability THEN    'Perform crossover.
  CALL CrossOver(Mate1, Mate2, NewIndividual)
ELSE    'Don't perform crossover.
  CALL NoCrossover(Mate1, Mate2, NewIndividual)    'Don't perform
                                                    crossover.
END IF
```

Figure 5.2. The main program (from previous page).

```
NEXT NewIndividual

CALL Mutate    'Mutate the temporary population.

CALL Replace    'Replace the old population completely by the new one.

CALL FindUnknowns    'De-code the new population to integers then real numbers.

CALL FindFitness    'Find the fitness of each member of the population.

CALL Statistics(MeanFitness, SumFitness, FittestIndividual, Ξ
              Generation)    'Find the mean fitness and the fittest individual.

CALL PrintGeneration(Generation, MeanFitness,
FittestIndividual)    'Print generation to file.

CALL Scaling(ScalingConstant, FittestIndividual, SumFitness, Ξ
MeanFitness)    'If linear fitness scaling is "on" then scale population prior to selection.

NEXT Generation    'Process the next generation.

CLOSE    'Close all files
```

Figure 5.2 (conclusion). The main program.

GENETIC AND OTHER OPERATORS

Figure 5.2 is the heart of the program: generations are cycled through while crossover and mutation build increasingly good solutions. In some ways the other operators are just detail; much can be learned by simply adjusting the internal GA settings (P_c, P_m, G etc.) and trying the program on various simple problems of your own choosing. However, an idea of how the genetic and other operators are implemented is probably necessary if the advanced techniques described in Chapter 4 are to be added.

In the following, each operation and its associated subroutine is discussed in turn.

DEFINING THE RANGE OF THE UNKNOWNS

Different problems will have unknown parameters being hunted between different bounds. The range of each unknown is set in DefineRange using the array Range. In Figure 5.3 the bounds of a two-dimensional problem are established; the list can easily be continued for higher dimensional problems after `NumberOfUnknowns` has been adjusted in the main program.

```
SUB DefineRange
'Defines the upper and lower bounds of each unknown.
'Add other ranges using the same format if more than two unknowns in the problem.

Unknown = 1 'the first unknown.
Range(1, Unknown) = 0 'The lower bound.
Range(2, Unknown) = 1 'The upper bound.

Unknown = 2 'the second unknown.
Range(1, Unknown) = -3.14159
Range(2, Unknown) = 3.14159

'Add other ranges if more than two unknowns in your problem.

END SUB
```

Figure 5.3. Defining the range of the problem.

OPENING THE RESULTS FILES

Output is via two files: LGADOS.RES and LGADOS.ALL (Figure 5.4). LGADOS.RES lists the generation, g; fitness, f_{max}, of the highest performing individual; the average fitness of the generation, f_{ave}; and the unknowns r_k contained in the fittest individual. LGADOS.ALL lists g, f, r_k and the binary chromosome C for all individuals in all generations, and hence can be very large. The files are comma-separated and can be loaded into most spreadsheets for manipulation and plotting.

The files are overwritten each time the program is run. Therefore it is important that the results are copied to files with more unique names at the end of a run if data are not to be lost.

```
SUB OpenFiles
'Open result files. See Chapter 2 for a description of their contents.

OPEN "LGADOS.RES" FOR OUTPUT AS #1
OPEN "LGADOS.ALL" FOR OUTPUT AS #2

END SUB
```

Figure 5.4. The opening of the results files.

CREATING THE INITIAL POPULATION

The initial population is created by throwing a random series of 0's and 1's (Figure 5.5). RND is a BASIC function which returns a random decimal between 0 and 1; if RND > 0.5 then a 1 is placed into the string, otherwise a 0 is inserted.

```
SUB InitialPopulation
'Create the initial random population.

FOR Individual = 1 TO PopulationSize

  FOR Bit = 1 TO TotalStringLength
    IF RND > .5 THEN
      Strings(Individual, Bit) = 1
    ELSE
      Strings(Individual, Bit) = 0
    END IF
  NEXT Bit

NEXT Individual

CALL FindUnknowns  'Decode strings to real numbers.

END SUB
```

Figure 5.5. Creating the initial population.

SELECTION

The program uses fitness proportional selection with replacement via a roulette wheel analogy. A random number is thrown using the BASIC function RND and is multiplied by f_{sum} (Figure 5.6). The wheel is then spun and the individual fitnesses added together until the sum is greater than or equal to this product. The last individual to be added is then the selected individual.

```
SUB Selection (Mate, SumFitness, MeanFitness)
```
'Select a single individual by fitness proportional selection.

```
Sum = 0
Individual = 0

RouletteWheel = RND * SumFitness

DO
   Individual = Individual + 1
   Sum = Sum + Fitness(Individual)
LOOP UNTIL Sum >= RouletteWheel OR Individual = PopulationSize

Mate = Individual

END SUB
```

Figure 5.6. Selection.

CROSSOVER AND NO CROSSOVER

A random number is used in the main program to decide whether the two mates (or parents) are passed to either CrossOver or NoCrossover. If the random number is less than or equal to P_c then crossover is used to build the child strings, which are passed to the new temporary population NewStrings (Figure 5.7). If not, the children are clones of their respective parents (Figure 5.8).

```
SUB CrossOver (Mate1, Mate2, NewIndividual)
```
'Perform single point crossover.

```
CrossSite = INT((TotalStringLength - 1) * RND + 1)  'Pick the cross-site
                                                       at random.

FOR Bit = 1 TO CrossSite 'Swap bits to the left of the cross-site.
   NewStrings(NewIndividual, Bit) = Strings(Mate1, Bit)
   NewStrings(NewIndividual + 1, Bit) = Strings(Mate2, Bit)
NEXT Bit

FOR Bit = CrossSite + 1 TO TotalStringLength 'Swap bits to the right of
                                                the cross-site.
```

Figure 5.7. Crossover being used to build two new members of the temporary population (continued over).

```
  NewStrings(NewIndividual, Bit) = Strings(Mate2, Bit)
  NewStrings(NewIndividual + 1, Bit) = Strings(Mate1, Bit)
NEXT Bit

END SUB
```

Figure 5.7 (conclusion). Crossover being used to build two new members of the temporary population.

```
SUB NoCrossover (Mate1, Mate2, NewIndividual)
'Pass the selected strings to the temporary population without applying crossover.

FOR Bit = 1 TO TotalStringLength
  NewStrings(NewIndividual, Bit) = Strings(Mate1, Bit)
  NewStrings(NewIndividual + 1, Bit) = Strings(Mate2, Bit)
NEXT Bit

END SUB
```

Figure 5.8. Cloning the parents to build two new strings.

MUTATION

Mutation is applied by calling subroutine Mutate (Figure 5.9) to step through the whole temporary population, visiting every bit in each string and throwing a random number. If this number is less than or equal to P_m the value of the bit is flipped.

```
SUB Mutate
'Visit each bit of each string very occasionally flipping a "1" to a "0" or vice versa.

FOR Individual = 1 TO PopulationSize
  FOR Bit = 1 TO TotalStringLength

    'Throw a random number and see if it is less than or equal to the mutation probability.
    IF RND <= MutationProbability THEN

      'Mutate.
      IF NewStrings(Individual, Bit) = 1 THEN
            NewStrings(Individual, Bit) = 0
      ELSE
            NewStrings(Individual, Bit) = 1
      END IF
```

Figure 5.9. Mutation (continued over).

```
        END IF

    NEXT Bit

NEXT Individual

END SUB
```

Figure 5.9 (conclusion). Mutation.

REPLACEMENT

The old population is replaced by the new one by copying NewStrings bit by bit into Strings and erasing the contents of NewStrings (Figure 5.10).

```
SUB Replace
'Replace the old population with the new one.

FOR Individual = 1 TO PopulationSize
   FOR Bit = 1 TO TotalStringLength
      Strings(Individual, Bit) = NewStrings(Individual, Bit)
   NEXT Bit
NEXT Individual

ERASE NewStrings   'Clear the old array of strings.

END SUB
```

Figure 5.10. Replacing the old population with the new one.

DECODING THE UNKNOWNS

The binary strings are converted to real-valued parameters for testing in the problem by calling FindUnknowns (Figure 5.11). This starts by calling FindIntegers to convert the strings to base-10 integers (Figure 5.12). These integers are then converted to real-valued parameters using the transformation of Chapter 2.

```
SUB FindUnknowns
'Decode the strings to real numbers.

CALL FindIntegers  'First decode the strings to sets of decimal integers.

'Now convert these integers to reals.
FOR Individual = 1 TO PopulationSize
  FOR Unknown = 1 TO NumberOfUnknowns
    Unknowns(Individual, Unknown) = Range(1, Unknown) + Ξ
Integers(Individual, Unknown) * (Range(2, Unknown) - Range(1, Ξ
Unknown)) / (2 ^ SubstringLength - 1)
  NEXT Unknown
NEXT Individual

END SUB
```

Figure 5.11. FindUnknowns calls FindIntegers and then converts the integers to reals within the bounds (range) of the problem.

```
SUB FindIntegers
'Decode the strings to sets of decimal integers.

DIM bit AS INTEGER

FOR Individual = 1 TO PopulationSize
  bit = TotalStringLength + 1
  FOR Unknown = NumberOfUnknowns TO 1 STEP -1
    Integers(Individual, Unknown) = 0
    FOR StringBit = 1 TO SubstringLength
      bit = bit - 1
      IF Strings(Individual, bit) = 1 THEN
        Integers(Individual, Unknown) = Integers(Individual, Ξ
                          Unknown) + 2 ^ (StringBit - 1)
      END IF
    NEXT StringBit
  NEXT Unknown
NEXT Individual

END SUB
```

Figure 5.12. Converting the strings to base-10 integers.

ASSIGNING THE FITNESS

The parameters extracted by FindUnknowns are tested as solutions to the problem at hand in FindFitness. Figure 5.13 shows code for the two-dimensional maximisation problem $f = x^2 + sin(y)$. The subroutine can be easily adapted to solve other problems. It is important to ensure that no negative fitnesses are assigned.

```
SUB FindFitness
'The problem at hand is used to assign a positive (or zero) fitness to each individual in turn.

'The problem is f = x^2 + sin(y).
FOR Individual = 1 TO PopulationSize
  Fitness(Individual) = Unknowns(Individual, 1) ^ 2 + Ξ
                               SIN(Unknowns(Individual, 2))
  If Fitness(Individual) < 0 then Fitness(Individual) = 0
NEXT Individual

END SUB
```

Figure 5.13. Inserting the unknowns into the problem and assigning a fitness to each member of the population in turn.

LINEAR FITNESS SCALING

For many problems, using some form of fitness scaling will greatly improve performance. Scaling (Figure 5.14) applies linear fitness scaling to the problem, with the amount of scaling being controlled by ScalingConstant. To switch scaling off, ScalingConstant can be set to zero within the constant declaration area of the main program. Scaling is applied just before selection. Because the results are output before this, the fitness reported is the true fitness, not the scaled fitness.

```
SUB Scaling (ScalingConstant, FittestIndividual, SumFitness,
MeanFitness)
'Apply Linear Fitness Scaling,
'    scaledfitness = a * fitness + b.
'Subject to,
'    meanscaledfitness = meanfitness
'and
'    bestscaledfitness = c * meanfitness,
'where c, the scaling constant, is set by the user.
```

Figure 5.14. Applying linear fitness scaling to the population prior to selection (continued over).

'If the scaling constant is set to zero, or all individuals have the same fitness, scaling is not applied.

```
IF ScalingConstant <> 0 AND Fitness(FittestIndividual) - Ξ
                                        MeanFitness > 0 THEN
```

  'Find a and b.

```
  a = (ScalingConstant - 1) * MeanFitness / Ξ
                  (Fitness(FittestIndividual) - MeanFitness)

  b = (1 - a) * MeanFitness
```

  'Adjust the fitness of all members of the population.
```
  SumFitness = 0
  FOR Individual = 1 TO PopulationSize
    Fitness(Individual) = a * Fitness(Individual) + b
    IF Fitness(Individual) < 0 THEN Fitness(Individual) = 0
```
'Avoid negative values near the end of a run.
```
SumFitness = SumFitness + Fitness(Individual) 'Adjust the sum of all the
                                                        fitnesses.

  NEXT Individual
```

  'Adjust the mean of all the fitnesses.
```
  MeanFitness = SumFitness / PopulationSize
END IF

END SUB
```

Figure 5.14 (conclusion). Applying linear fitness scaling to the population prior to selection.

ELITISM

Elitism is applied by checking if the fittest individual has a lower fitness than the elite member of the last population; if so, a randomly selected individual is replaced by the old elite member (Figure 5.15).

```
SUB Elite (SumFitness, FittestIndividual)
```
'Applies elitism by replacing a randomly chosen individual by the elite member
'from the previous population if the new max fitness is less then the previous value.

```
IF Fitness(FittestIndividual) < EliteFitness THEN

  Individual = INT(PopulationSize * RND + 1) 'Chosen individual to be
                                                        replaced.
```

Figure 5.15. Elitism (continued over).

```
     FOR Bit = 1 TO TotalStringLength
        Strings(Individual, Bit) = EliteString(Bit)
     NEXT Bit

     Fitness(Individual) = EliteFitness

     FOR Unknown = 1 TO NumberOfUnknowns
        Integers(Individual, Unknown) = EliteIntegers(Unknown)
        Unknowns(Individual, Unknown) = EliteUnknowns(Unknown)
     NEXT Unknown

     FittestIndividual = Individual

END IF

FOR Bit = 1 TO TotalStringLength
        EliteString(Bit) = Strings(FittestIndividual, Bit)
NEXT Bit

EliteFitness = Fitness(FittestIndividual)

FOR Unknown = 1 TO NumberOfUnknowns
   EliteIntegers(Unknown) = Integers(FittestIndividual, Unknown)
   EliteUnknowns(Unknown) = Unknowns(FittestIndividual, Unknown)
NEXT Unknown

END SUB
```

Figure 5.15 (conclusion). Elitism.

STATISTICS

Subroutine statistics (Figure 5.16) is used to find f_{ave} (MeanFitness), f_{sum} (SumFitness) and f_{max} (MaxFitness). These are required if elitism is being used and are also reported by PrintGeneration. The subroutine also calls Elite if required.

```
SUB Statistics (MeanFitness, SumFitness, FittestIndividual, Ξ
                                                   Generation)
'Calculate the sum of fitness across the population and find the best individual,
'then apply elitism if required.

FittestIndividual = 0
MaxFitness = 0

FOR Individual = 1 TO PopulationSize
  IF Fitness(Individual) > MaxFitness THEN
    MaxFitness = Fitness(Individual)
    FittestIndividual = Individual
  END IF
NEXT Individual

IF Elitism = "on" THEN 'Apply elitism.
  CALL Elite(SumFitness, FittestIndividual)
END IF

SumFitness = 0 'Sum the fitness.
FOR Individual = 1 TO PopulationSize
  SumFitness = SumFitness + Fitness(Individual)
NEXT Individual

'Find the average fitness of the population.
MeanFitness = SumFitness / PopulationSize

END SUB
```

Figure 5.16. Finding f_{ave}, f_{sum} and f_{max} and calling Elite.

PRINTING THE RESULTS

PrintGeneration (Figure 5.17) is used to output a single generation to the screen and to file. The file LGADOS.ALL contains all individuals from all generations and hence can be rather large. If the information it contains is not required, the corresponding lines should be commented out. BASIC uses a semicolon at the end of a PRINT line to stop a carriage return from being automatically included. The files are comma separated to allow easy importing of the results into a spreadsheet or plotting package.

```
SUB PrintGeneration (Generation, MeanFitness,
FittestIndividual)
'Print results to the screen and the files.

PRINT Generation; Fitness(FittestIndividual); MeanFitness;
                                                    'Screen.
PRINT #1, Generation; ","; Fitness(FittestIndividual); ","; Ξ
                                MeanFitness; 'File LGADOS.RES.

FOR Unknown = 1 TO NumberOfUnknowns
   PRINT Unknowns(FittestIndividual, Unknown); 'Screen.
   PRINT #1, ","; Unknowns(FittestIndividual, Unknown); 'File
                                                  LGADOS.RES
NEXT Unknown

PRINT 'Carriage return.
PRINT #1, 'Carriage return.

FOR Individual = 1 TO PopulationSize

   PRINT #2, Generation; ","; Fitness(Individual); ","; 'File
                                                  LGADOS.ALL

   FOR Unknown = 1 TO NumberOfUnknowns
      PRINT #2, Unknowns(Individual, Unknown); ","; 'File LGADOS.ALL
   NEXT Unknown

   FOR Bit = 1 TO TotalStringLength
      PRINT #2, RIGHT$(STR$(Strings(Individual, Bit)), 1);","; 'File
                                                  LGADOS.ALL
   NEXT Bit

   PRINT #2, 'Carriage return

NEXT Individual

END SUB
```

Figure 5.17. Printing the results from a single generation to the screen and the two output files.

RUNNING AND ALTERING THE PROGRAM

After starting QBASIC, load LGADOS.BAS using the FILE menu and run the program using the RUN menu. Then open the two results files and familiarise yourself with the contents. To obtain meaningful answers to many questions (such as the best value of P_m for a particular problem) it is imperative that multiple runs of the program are used. Because the program always uses the same filenames for the results files, all previous results are lost each time a new run is embarked upon. It is therefore important to change the name of these files at the end of run. (An alternative would be to make the necessary changes to the program to allow the user to input their own filenames.)

It is relatively straightforward to adapt the program to solve other problems. This involves three main changes (which should only carried out on a copy of the program):

1. Within the "set constants" area of the main program adjust the values of any constants that need changing;
2. Press F2 to access the subroutine list and edit `DefineRange` to reflect both the number of unknowns in the problem and their ranges; and
3. Edit `FindFitness` to reflect the new problem.

Adaptations to `FindFitness` can be made either by inserting the required code directly into the program or, if using a compiler, by "chaining" to another compiled program (by using the CHAIN statement) and using files (or other methods) to transfer the solution vectors and the fitness between the two programs.

CHAPTER 6

APPLICATIONS OF GENETIC ALGORITHMS

A quick search of a publication database shows that the number of GA applications is growing on a daily basis (Figure 6.0.1). The range of applications is truly stunning, including evolving computer programs [KO92,KO94], estimations of protein structure [SC92] and designing water networks [HA97].

The eight applications given in this book have been chosen for several reasons: they are all *practical* applications in science or engineering; all except one uses relatively simple binary-encoded GAs of the type described in the earlier chapters; further, they complement rather than echo the applications given in other texts (discussed below). Although the requirement that the examples use only binary representations and no techniques not covered in the earlier parts of the text has somewhat limited the range of applications that could be presented, this has been done to forcefully demonstrate that relatively simple algorithms can help in solving complex problems. Each application attempts to illustrate some of the difficulties in applying GAs to various real-world problem domains (Table 6.0.1). They range from the simple estimation from experimental data of five real-valued unknowns, to the use of a messy GA to solve a complex water network problem.

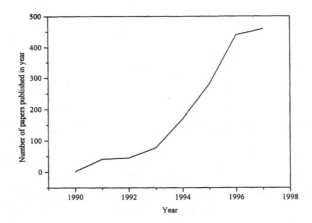

Figure 6.0.1. The continuing growth in the number of papers using GAs published each year indicates a burgeoning acceptance of the technique. (Data obtained by searching for the string "genetic algorithm" within the *scientific* publications database at the University of Bath; papers within other fields are therefore not included.)

§	Application	Main attributes
6.1	Image Registration	• use of a simple GA (LGA); • at most 25 real-valued unknowns; • use of an approximate fitness function which grows in accuracy with time.
6.2	A Simple Application: Recursive Prediction of Natural Light Levels	• GA applied in a recursive manner to a fitness landscape that is not static, i.e. the algorithm is hunting a noisy moving target and two identical, but time separated, genotypes may not give rise to identical values of fitness; • a simple least-squares problem comparing an experimental data set with a simple model; • few unknowns and limited accuracy required; and • (therefore) a short string length.

§	Application	Main attributes
6.3	Water Network Design	• discrete unknowns; • Gray coding; • linear rank selection; • uniform crossover; • a penalty function to handle unfeasible solutions; and • the requirement for only a near-optimal, but robust, solution.
6.4	Ground-State Energy of the ±J Spin Glass	• long genotypes, $L > 3000$; • large number of Boolean unknowns, $M > 3000$; • inclusion of an additional local search heuristic (directed mutation); and • direct correspondence between the problem-space and the string representation, negating the need to encode the unknowns.
6.5	Estimation of the Optical Parameters of Liquid Crystals	• the development of a system to replace a previously human-guided search; • use of a series of sequential adaptations to a simple GA; and • the inclusion of a traditional search routine to create a hybrid algorithm.
6.6	Design of Energy Efficient Buildings	• a range of different variable types (binary, integer and real); • the need for a diverse range of approximate solutions together with the global optimum; • the need for human-based final selection; and • the need to avoid re-estimating the fitness of any already processed design. • use of remainder stochastic sampling and a generation gap

§	Application	Main attributes
6.7	Human Judgement as the Fitness Function	• a GA being driven by aesthetic judgement, rather than a numerical measure of fitness; • a problem where identical genotypes may be given non-identical values of fitness; and • a system where relatively few fitness evaluations are possible.
6.8	Multi-objective Network Rehabilitation by Messy GA	• a sparse problem space; • use of a *messy* algorithm; • a multi-objective problem; and • fitness sharing.

Table 6.0.1 the various problems detailed in this chapter and their main attributes.

As with any book, the number of applications is limited by spatial constraints. If none of them match your own areas of interest, the following introductory texts should be studied for further applications and references. Alternatively, an electronic search should be made of a publication database.

Davis [DA91]
• Parametric design of aircraft
• Routing in communications networks
• Robot trajectories
• Nonlinear systems and models of international security
• Strategy acquisition
• Artificial neural networks
• Mineral separation by air-injected hydrocyclone
• Fault diagnosis
• Conformational analysis of DNA
• Interpretation of sonar signals
• Engineering design
• Travelling salesman problem
• Scheduling

Goldberg [GO89]
• Optimisation of engineering structures
• Iterated prisoner's dilemma

- Machine learning

Michalewicz [MI94]
- The transportation problem
- Travelling salesman problem
- Graph drawing
- Scheduling
- Partitioning
- Robotic path planning
- Machine learning

Mitchell [MI95]
- Evolving computer programs in LISP
- Cellular automata
- Time series analysis
- Protein structure
- Evolving weights, learning rules and architectures for artificial neural networks
- Use of GAs in modelling evolution, learning, sexual selection and ecosystems.

Zalzala and Fleming [ZA97]
- Control systems
- GAs and fuzzy logic
- Artificial neural networks
- Chaotic system identification
- Job shop scheduling
- Robotic systems
- Aerodynamics
- VLSI layout

6.1 IMAGE REGISTRATION

Genetic algorithms have a long history of use in image registration and in particular with medical image registration [for example FI84]. This is not surprising as one can easily imagine that many such problems will contain search spaces littered with local minima. Unlike x-rays, modern medical imaging techniques also have the distinction of generating three-dimensional visualisations of the human body, greatly increasing the scale of the search space.

The work reported here is taken from work by Coley and Bunting [CO94] and Wanschura, Coley, Vennart and Gandy, and attempts to register a pair of time-separated magnetic resonance imaging (MRI) data sets. (This time separation may be over several months and with the images recorded on differing machines.) The technique is useful for monitoring small changes in anatomy, such as disease progression with time (e.g. degradation of cartilage in arthritic disease), pre/post-surgical studies and signal intensity (e.g. functional brain imaging) in otherwise identical images. It provides a simple application of a GA to a problem requiring:

- the use of a simple GA (LGA);
- at most 25 real-valued unknowns; and
- (because of the large amount of experimental data) the use of an approximate fitness function which grows in accuracy with generation.

INTRODUCTION

Even when taken with registration in mind, time separated MRI data sets are unlikely to be perfectly registered. Misalignment of the object under investigation within the RF coil, small anatomical changes and changes in system settings will all affect the registration of the images. This misalignment is unlikely to be of concern if the images are presented as planar slices through the objects and the time-separated sets are kept distinct. However, if an accurate determination of very small changes is required, the images will require to be registered before further operations (e.g. subtraction) are performed. The approach adopted for identifying changes by registration is based simply on a calculation of pixel by pixel intensity difference followed by subtraction.

To obtain the registration the first image must be transformed (translated, rotated or deformed) until the best match with the second image is found. The similarity of the two images can then be quantified in a cost function, Ω. By using a numerical optimisation method to find the minimal cost (or maximal similarity) a fully automatic procedure to extract the region of interest is produced.

MRI data sets typically contain in excess of 16,000 points per image, and the search space for implementing registration is likely to contain a large number of local optima which do not give the correct alignment of the images. Although the GA has been promoted as an aid to the registration of medical images taken by different techniques (e.g. MRI together with X-ray) [FI84,HI94,HI96], little work has been carried out to indicate the accuracy of the method when used on typical pairs of time-separated MRI data sets [CO94,CO96]. In particular, the abilities of the method have not been studied for sets misregistered by amounts typical of clinical scans. (The data sets used in this study were collected for other general purposes, and not with registration in mind).

METHOD

The two image data sets were stored as greyscale (0-255) intensities of 128 x 128 pixels. Automatic registration of the two images requires (a) a transformation and (b) a measure of similarity. The transformation applied is given by the following:

$$\bar{x} = a_0 + a_1 x + a_2 y + a_3 z + a_4 xy + a_5 xz + a_6 yz + a_7 xyz \ , \tag{6.1.1a}$$

$$\bar{y} = a_8 + a_9 x + a_{10} y + a_{11} z + a_{12} xy + a_{13} xz + a_{14} yz + a_{15} xyz \tag{6.1.1b}$$

and

$$\bar{z} = a_{16} + a_{17} x + a_{18} y + a_{19} z + a_{20} xy + a_{21} xz + a_{22} yz + a_{23} xyz \ , \tag{6.1.1c}$$

where x, y, z are the three Cartesian spatial dimensions and the a_i are the adjustable parameters whose values we are trying to estimate ($z = \bar{z} = 0$ in the 2-D case, leaving eight unknowns). The transformation can be visualised as a rubber cube which allows the image to be translated, re-scaled, sheared or rotated. For example, a_0, a_8 and a_{16} describe the translation of the image in the x-, y- and z-directions respectively.

The measure of fitness, or success of the transformation is based simply the point-by-point absolute difference between the two images:

$$f = A - \frac{1}{m} \sum_{j=1}^{m} |\Delta c(j)| = A - \Omega \ , \tag{6.1.2}$$

where m = number of points considered, Δc = colour (or greyscale intensity) difference between the same point in the first and the transformed second image, and A is a constant to convert the minimisation to a maximisation. (More complex measures have been suggested [TO77,BR91].) Because the two images may well have different overall intensity distributions (due to the way the imaging machines were staffed and operated—with the time separation possibly being over several months) an additional unknown parameter, a_{24}, is required to equalised the distributions:

$$\Delta c(j) = c_1(j) - a_{24}c_2(j)$$

where $c_1(j)$ and $c_2(j)$ are the individual colours (or greyscale intensities) of pixel j for the two images. (Note, a_{24} is the same for all j.)

APPROXIMATE FITNESS ESTIMATION

To speed calculation, the points j must be kept to a minimum. Initial filtering is carried out by excluding all points with near zero intensities. Most of these points represent the background black outside the anatomical structure being studied. This typically halves the number of points processed. Pixels j are selected as a random subset of the complete data (this subset must be large enough to contain a statistical representation of the image, initially 2-3%), with the size of this subset growing as convergence is reached [FI84]. It is this statistical approach which is responsible for much of the efficacy of the approach. Because of the high levels of noise within the images, much of the intensity of a pixel within this subset might simply be stochastic in nature. To alleviate this, the intensities of pixels within the images were averaged with the surrounding ones. For $g = 1$, and 2-D images, this smoothing took place over 25 pixels centred on the pixel in question, the size of the grid reducing to unity by $g = G$.

LGA was used with $N = 50$ initial guesses, $P_c = 0.8$ and $P_m = 0.001$. The registration obtained by the method can be demonstrated by comparing slices from subtractions of the two sets, with and without a transformation being applied.

In order to test the accuracy of the approach outlined above, the technique was initially applied to the problem of registering two time-separated images (Figure 6.1.1a, average pixel intensity, $I_{av} = 49.8$, $\sigma = 45.9$ and Figure 6.1.1b, $I_{av} = 53.4$, $\sigma = 50.3$) of the right-hand index distal-interphalangeal (DIP) joint of a healthy volunteer.

3-D images with 200μm in-plane resolution were collected with 16 contiguous slices of 1 mm thickness. After the initial scan, the finger was removed from the RF coil, then repositioned within the coil ready for the second imaging experiment. (This repositioning mechanism has been found to be inherently inaccurate, and can often result in misalignment of the joint by the order of mm.) A second set of images was taken, and upon completion, the central image slice from each data set was selected, and the GA applied to align the two sets.

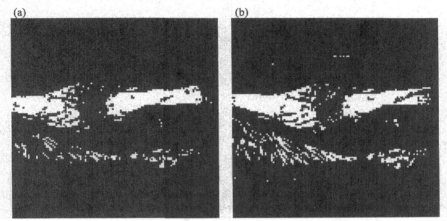

Figure 6.1.1. (a) Planar slice through image set 1: distal-interphalangeal joint in sagittal section. (b) Equivalent planar slice through image set 2. Note, the finger is displaced by approximately 2 mm. (The figures are binary versions of the 256 grey-scale intensities used.)

RESULTS

The generation, or time, evolution of the fitness is displayed in Figure 6.1.2. The solution is seen to have converged after approximately thirty generations. This point took approximately 20 s to reach on a personal computer, demonstrating the efficiency of the algorithm—which was implemented in PASCAL.

122

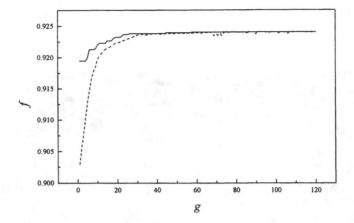

Figure 6.1.2. The time evolution of the fitness; the highest fitness within any generation is shown as a solid line, the mean fitness of the population as a dashed line.

The degree of misalignment reduction can be viewed by comparing subtractions of the data sets with, and without, the transformation. Subtractions of correctly registered, identical, sets should result in totally blank images. Subtracting the two images given in Figure 6.1.1, without first applying a transformation, results in Figure 6.1.3a ($I_{av} = 25.4$, $\sigma = 30.9$). The "ghost" image seen is typical of misalignment. The only enhancement to the final image is noise reduction by rejection of very low intensity pixels. For comparison an image obtained by identical subtraction, but after application of the transformation generated by the GA, is given in Figure 6.1.3b ($I_{av} = 2.1$, $\sigma = 11.1$). The ability of the GA to accurately align the images is clearly shown by there being almost no structures in the image, and no ghost image (also implying that there are no anatomical differences between the two data sets); what remains is essentially noise. A more quantitative measure of alignment, apart from the much reduced values of I_{av} and σ, is the number of fully connected pixels, P_f, which reflects the information content of the image. A pixel is fully connected if, and only if, it has a non-zero intensity and its eight nearest neighbours also have non-zero intensities. If the two images have been successfully aligned to the image resolution (1 pixel, or 200μm) then subtraction will give $P_f \approx 0$. For Figure 6.1.1a and 6.1.1b, $P_f = 15793$ and

15723 respectively; for Figure 6.1.3a $P_f = 12907$, and for Figure 6.1.3b, $P_f = 0$ indicating registration.

The technique is useful for monitoring small changes in anatomy, such as disease progression with time (e.g. degradation of cartilage in arthritic disease), pre/post-surgical studies and signal intensity (e.g. functional brain imaging) in otherwise identical images.

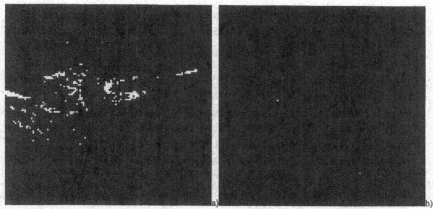

Figure 6.1.3a: Planar slice after the subtraction of the images without the GA-based registration. The ghost image of the joint is due to misalignment of the data sets. Figure 6.1.3b: Planar slice through the subtraction of the images after the GA-based registration; almost all misalignment has been removed. (Note, these figures are binary versions of the 256 grey-scale intensities used.)

6.2 RECURSIVE PREDICTION OF NATURAL LIGHT LEVELS

Controlling artificial lights within buildings to act solely as a supplement to available daylighting requires continuous knowledge of natural lighting levels within each room. Although this information is readily obtained by measurement whilst lights are extinguished, once illuminated the determination of the underlying natural light level is not so straightforward. This application describes the use of a genetic algorithm as the heart of a self-commissioning, adaptive algorithm capable of the real-time prediction of natural light levels at chosen points within a room using external measurements of vertical and horizontal plane illuminance. Such a system forms the basis of a robust and practical lighting controller.

This application (based on reference [CO97]) is very simple, requiring only the identification of five unknowns—the accuracy of which is not particularly important. The GA is applied in a recursive manner: as new sets of data arrive the algorithm is re-run, without re-initialisation, for a fixed number of generations. The results are compared with those found by a traditional method [CO94a]. It illustrates the following:

- a binary coded GA applied in a recursive manner to a fitness landscape that is not static, i.e. the algorithm is hunting a moving target and two identical, but time-separated, genotypes may not give rise to identical values of fitness;
- fitness-proportional selection;
- single point crossover;
- a simple least-squares problem comparing an experimental data set with a simple model;
- a search space known to contain local optima;
- many equally possible, often distant, solutions;
- few generations to convergence;
- few unknowns;
- limited accuracy required; and
- (therefore) a short string length.

INTRODUCTION

This application discusses the employment of a GA to the problem of predicting the natural horizontal plane illuminance (i.e. that falling on a desk or worksurface) within a room, from measurements of illuminance outside the building envelope and at some point distant. Attempts at such predictions have been made before [CO94a] using a recursive least squares algorithm, allowing results for a GA to be contrasted with those of a traditional algorithm.

Lighting is often the largest single item of energy cost in offices and considerable savings can be made [EE91] implying reductions in both cost and resultant emissions of gases implicated in global warming. Savings can be effected by a combination of higher efficiency lamps, more efficient fittings, better controls and increased usage of daylighting. Here the approach is a novel lighting controller which enables maximum use to be made of available daylight.

The availability of high levels of natural light within a space is not, of itself, sufficient to ensure that less use will be made of artificial lighting. For manually controlled systems, studies have shown [HU79] that it is the perceived level of light on initially entering the room which chiefly determines whether lights will be illuminated. Once switched on, it is very unlikely that they will be switched off—notwithstanding any subsequent increase in natural light levels. Automatic photocell-based controllers face a similar difficulty. It is straightforward for such systems to prevent the initial use of artificial lights if levels of natural light measured at internal sensors are high enough to obviate their use. Once illuminated, however, it is more difficult to decide whether the underlying level of daylighting is sufficient to enable them to be extinguished. There are various solutions to this problem.

One apparently attractive approach to the problem is to make measurements of external lighting levels and to relate these to expected internal levels. This is far from easy to accomplish. The relationship between internal and external illuminance is time-dependent on an hourly and seasonal scale. Thus any attempt to determine a simple ratio which relates the two is unlikely to be successful. A better approach would be to relate external horizontal and vertical plane illuminance, in a number of directions, to internal horizontal plane illuminance [HA88]. The advantage is that vertical plane illuminances contain directional, and therefore time, information. A means must still be found, however, to determine the ratios which relate the various components of vertical plane illuminance to internal conditions. This can be done by calculation for each space, but there is a better approach, which also allows such ratios to be time-dependent.

Experience with lumped-parameter thermal models [CR87,CR87a] led to the development of a computer routine for the real-time identification of the model parameters from the measured building response to energy inputs [PE90,PE90a,CO92]. There is obvious potential here for application to the lighting problem just discussed. Given a simple model relating internal horizontal plane illuminance to the separate components of the external horizontal and vertical plane illuminance, a similar technique could be used to identify the unknown parameters as employed in the thermal case. Such an approach was successfully tried [CO94a] by using a recursive least squares algorithm. However, although the *average* error was found to be small, the largest errors in prediction (which were possibly systematic) occurred during times of high illuminance, just when any practical controller should be signalling for the artificial light to be extinguished. It was unclear whether this poor performance was due to the model or the algorithm used. It was decided to see if better predictions could be made by using a GA, thus hopefully lending support to the model. If indeed the GA did perform better, this would probably imply it had successfully navigated around one or more local optima to a better solution.

THE MODEL

The model chosen to relate the internal horizontal plane illuminance I_m^{in}, at time t, at a given point in any room m in the building, to the five external measured vertical and horizontal plane illuminances, I_j^{ex}, falling on the roof of the building, was the simple parameterisation,

$$I_m^{in}(t) = \sum_{j=1}^{5} b_m^j I_j^{ex} \; , \tag{6.2.1}$$

where the b_m^j are the numerical attenuation coefficients which contain information on, for example, the attenuating power of the windows and the reflectivity of internal and external surfaces.

It is worth noting that the model described here does not need any information on the alignment of the external sensors, nor does it require knowledge of the orientation of the windows. This will allow any system that is developed from this model to be largely self-commissioning.

Equation (6.2.1) is a somewhat arbitrary parameterisation of the problem. Thus the b_m^j will not, in general, form a linearly independent set. However this should not be a problem as there is little need to find a unique set $\{b\}$, but only any set in the parametric space which is capable of representing the response of the system. This will only hold true if no attempt is made to place a precise, physical meaning on the individual elements of $\{b\}$, but instead to regard them solely as parameters. It is known [CO94a] that the space contains many local optima and that there are many sets $\{b\}$ able to provide solutions of various accuracy.

PARAMETER EXTRACTION

For any particular room, (6.2.1) is of the form

$$x_o = \sum_{j=1}^{5} a_j x_j \; . \tag{6.2.2}$$

Equation (6.2.2) is the form needed for the classical linear least-squares problem in which a set of unknown parameters $\{a\}$ describing an observed system is to be determined from k measurements at different times of a set of observables $\{x\}$.

In the linear least-squares problem it is assumed that the k successive measurements of the set x can be represented by

$$x_{oi} = \sum_{j=1}^{N} a_j x_{ji} + \varepsilon_i .$$

(6.2.3)

where $i = 1, 2 \ldots k$; ε_i is an error term, and in this case $N = 5$.

Although strictly speaking the ε_i should be Gaussian and serially uncorrelated, least-squares gives reasonable results with most error distributions encountered in practice. Specifically, estimates $\hat{a}_j$ of the unknown parameters a_j are chosen to minimise the function Ω, where

$$\Omega = \sum_{i=1}^{k} \left(\sum_{j=1}^{N} x_{ji} \hat{a}_j - x_{0i} \right)^2 .$$

(6.2.4)

In order to solve (6.2.4) using a GA, the five unknown parameters represented by the elements of $\{\hat{a}\}$ were encoded as binary strings of length 10. Using the knowledge gained from reference [CO94a] it was decided to limit the search range to ±100. Thus the search space is discretised to an accuracy of better than one part in a thousand.

SECULAR TRENDS

In (6.2.4) the sum runs over all values of i, i.e. all the collected data. This has the advantage of smoothing fluctuations in the data caused by inaccuracies in the measurements (caused, for example, by the temporary use of blackouts, or wildlife interfering with the external sensor), thereby stabilising the values stored in $\{\hat{a}\}$, and making the algorithm robust. However, although stability against fluctuations in the data is a good thing, there is a desire to track secular trends in the response characteristics of the system: for example, the dirtying of windows or the construction of a nearby building. This problem can be solved by various methods [YO74]: for example, by using an exponentially weighted past averaging method to curtail the memory of the estimation procedure in least-squares analysis. In systems where the variation of the parameters is known *a priori*, more advanced methods can be used, for example the Kalman filter estimator [KA60]. As very little *a priori* information of the nature of the time variation of the parameters has been assumed, a simpler method is adopted. The sum in (6.2.4) is run over a sliding window of fixed width w. Thus the arrival of a new data point (the kth) causes the sum to run from $k-w$ to k.

As any trends are presumed to be slow, consecutive predictions of $\{\hat{a}\}$ are likely to be similar. Thus it would seem appropriate not to the reset the GA to a random initial population upon receipt of a new data point. The initial population is simply set to the final population from the last run.

APPLICATION

A computer code ILLUMIN was written to apply LGA to the model and calculate the five unknown parameters that characterise the response of a room.

The data analysed here were recorded at the Physics building of the University of Exeter during the spring of 1992. Five light-sensitive resistors were attached to the top and the four vertical faces of a small plastic cube. This sensor array was housed in a glass enclosure and placed on the roof of the building. Within a room in the same building a single sensor was set in a horizontal orientation to measure the internal light level. The time series of measurements from these sensors was monitored constantly and recorded on a data logger every ten minutes (Figure 6.2.1). Two consecutive sets were collected, one for training of the GA and the other for testing the results.

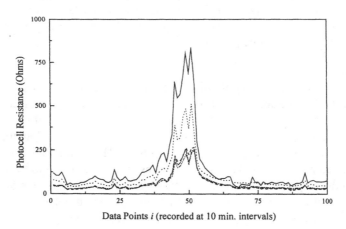

Figure 6.2.1. A short fraction of the time series of measurements from the five external sensors (nocturnal periods removed), with each line representing a different sensor.

RESULTS

Table 6.2.1 shows the parameter values identified by ILLUMIN using the first seven days' data, together with the values found by the recursive method. f^* was found after 24 generations for $N = 100$, $P_m = 0.005$, $P_c = 0.6$, $\acute{e} = 1$, fitness-proportional selection, and w set to the width of the data set. The minimisation of (6.2.4) was recast as a maximisation of fitness by setting $f = A - \Omega$, where the constant A is large enough to ensure f is positive. The identified values are similar, but not identical, to those produced by the traditional recursive method [PL50].

Parameter	Recursive	GA
a_1	12.156	11.7
a_2	0.5328	0.5
a_3	22.522	21.9
a_4	22.262	22. 1
a_5	-36.713	-35.8

Table 6.2.1. The final (dimensionless) parameter values found by recursive least squares [CO94a] and by the genetic algorithm, for the data of Figure 6.2.1.

The model represented by (6.2.2) and the fit represented by the parameter values given in Table 6.2.1 can be tested by using (6.2.2) to generate a time series of internal light levels, and comparing this with the second observed series (Figure 6.2.2). The results from the GA show exceptionally good agreement and a large improvement (particularly at higher light levels) over those from the classical recursive least-squares algorithm. The final parameter set is found to predict the internal light level from observed external levels to within an RMS error of 3% at an illuminance of 500 lux. This improvement implies that the GA has better navigational properties within the space. In particular, as both solutions are close, not just in terms of the result but the parameter set itself, it would seem that the GA was able to "jump" over one or more local optima to find a better solution. However, without an enumerative search of the space it is impossible to say if the values found represent the true global optimum.

In practice k extends indefinitely, and thus many totally separated sets of width w are collected so that the controller can take account of slow secular trends in the parameters (caused, for example, by seasonal transients in the reflectivity of the land surrounding the building and changes in tree foliage). This adaptive mode is also suitable for signalling excursions in the fitted attenuation parameters beyond pre-determined limits (caused, for example by the failure of the occupants to open curtains). The method has proved successful with $w = 600$ and the GA being re-run without re-initialisation for a further 24 generations for each subsequent set. An alternative approach might be to set $w = 1$ and only allow the algorithm to advance a single generation for each new data point. However, because of the possibilities of large errors in individual data points, due in this instance to environmental factors, the performance of the GA might well be compromised.

This opens the way to integrating such models with lighting controllers. Such a controller would employ this technique to maintain an updated estimate of the attenuation factors, $\{b\}$, during periods when artificial lights were not in use. Once the lights were illuminated, the controller would use the model represented by (6.2.1) to predict natural internal light levels from the continuing measurements of external light levels I^{ex}_j. If the predicted internal illuminance exceeded a predetermined threshold value (set by the occupants), the controller could then call for the artificial lights to be extinguished. Such a prototype has been used to control the illumination within an office space.

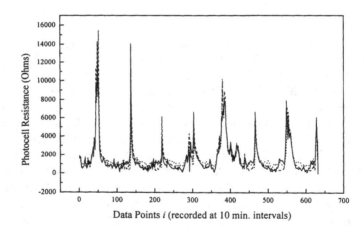

Figure 6.2.2. Comparison of measured and predicted light levels (nocturnal periods removed): GA solid line, recursive classical dotted line, and measured dashed line.

6.3 WATER NETWORK DESIGN

Such is the scale of infrastructural asset represented by a large water distribution network that the use of a sub-optimal design can have considerable cost implications. Unfortunately the complexity of such systems is such that the identification of the optimal design for a particular network is far from easy. By developing a GA-based design tool, GANET [SA97], for water distribution network planning and management, Godfrey Walters and Dragan Savic are attempting to rapidly identify near-optimal solutions.

In the work described here [SA97], GANET was applied to both illustrative and real water networks. The following features are exhibited:

- discrete unknowns;
- Gray coding;
- linear rank selection;
- uniform crossover;
- a penalty function to handle unfeasible solutions; and
- the requirement for only a near-optimal, but robust, solution.

The penalty function is particularly interesting because it is generation-dependent.

INTRODUCTION

The problem of selecting the optimal set of pipe diameters for a network so as to minimise cost has been shown to be an NP-hard problem[†] [YA84]. The simulation of the hydraulic behaviour of a single network itself is a difficult task, requiring a simultaneous consideration of energy and continuity equations and the head-loss function [WO93]. The actual network layout is often determined by such factors as the location of roads, leaving only pipe cost to be minimised. The solution itself is constrained by requirements of minimum flows and pressures at various points on the network. The pipe diameters themselves will not in general be continuous variables, but will be defined by the range of commercially available sizes.

The objective function, Ω, to be minimised is the cost, c, given by a function of pipe diameters Φ_i and fixed lengths λ_i only:

[†] NP hard problems are ones that cannot be solved in polynomial-time. Essentially this means that the complexity of the problem grows at a faster than polynomial rate as more unknowns are added, making all but the simplest problems impossible to solve.

$$\Omega = \sum_{i=1}^{N_p} c(\Phi_i, \lambda_i) \ , \tag{6.3.1}$$

where N_p is the total number of pipes in the network.

Ω requires minimising under several constraints. At each junction node the flow into the junction, Q_{in}, and the flow out of the junction, Q_{out}, are connected through the continuity constraint:

$$\sum Q_{in} - \sum Q_{out} = Q_e \ , \tag{6.3.2}$$

where Q_e describes the demand (or inflow) at the node. For each loop within the system energy conservation implies that:

$$\sum h_f - \sum E_p = 0 \ , \tag{6.3.3}$$

where E_p is the energy input by a pump and h_f is termed the *head-loss*. In order for the network to be useful a minimum head will be required at each node. These minimum heads H_j^{min} , at each node j form a set of constraints, H_j such that:

$$H_j \geq H_j^{min} \ ; j = 1, \ldots . N_n \ , \tag{6.3.4}$$

where N_n is the total number of nodes.

In order to confirm that any network designed by the GA satisfies these constraints requires the solution of a hydraulic model of the system. The most commonly used formula to describe the relationship between pressure drop, flow rate, pipe length and pipe diameter is the empirical Hazen-Williams (H-W) relationship [WA84]; in empirical units:

$$v = b R_h^{0.63} S_f^{0.54} \ ,$$

where v is the flow velocity, R_h is the hydraulic radius, S_f the hydraulic gradient, and b a dimensional coefficient. For a pipe of length λ and flow Q the head loss is given by:

$$h_f = \omega \frac{\lambda}{b^{1/0.54} \Phi^{2.63/0.54}} Q^{1/0.54} \ ,$$

where ω is another numerical conversion constant.

PREVIOUS SOLUTION TECHNIQUES

It has been suggested [YA84] that only explicit enumeration or an implicit numeration technique such as dynamic programming can guarantee finding the optimal solution. The problem is non-linear due to the energy constraints and further complicated by the requirement to use discrete-sized pipe diameters. The problem is NP-hard and intractable for even relatively small ($N_p < 20$) networks. However, as is typical with many real-world problems, the inability to guarantee an optimal solution is not of prime importance. The goal is the identification of the lowest-cost system given a realistic amount of time in which to do the calculation. Due to engineering and other factors which cannot be fully described in the model, the final solution implemented might only in part be described by the optimum identified by the modelling processes. Because of such considerations and the limited amount of time available to find a working solution, it is important that the algorithm identifies near-optimal solutions rapidly. Even if other methods are able to find better solutions given enough time, it is the performance of the method within a time constraint which is important. With a network of only 20 pipes and a set of 10 possible pipe diameters there are 10^{20} different possible designs. Clearly only a fraction of this search space can be examined by any practicable method.

Historically, much of the search space has been eliminated either by using a simplified approach [GE85,MU92], or by additional human intervention [WA85]. Other approaches include attempts to reduce the problem to a sequence of linear subproblems [SH68,FUJ90]. This can lead to split pipe designs, with one or two pipe segments of differing discrete sizes between nodes. In reality, such segments will need to be replaced by a single diameter before implementation.

Nonlinear programming techniques have been applied to the problem [EL85,DU90], but have treated pipe diameters as continuous variables and are limited to problems of small size.

Increasingly, evolutionary inspired programming methods have been applied to hydraulic networks [GO87a,WA93,MU92,WA93a]

GANET: A GA MODEL FOR LEAST-COST PIPE NETWORK DESIGN

The use of a GA to solve a network problem can be illustrated by using GANET to solve the two-loop network studied initially by Alperovits and Shamir [AL77] (Figure 6.3.1). In the following, the GA rather than the water engineering aspects will be concentrated upon. In particular, consideration will focus upon the hunt for a near-optimal solution, the use of Gray coding, the form of the unknowns, and the penalty function.

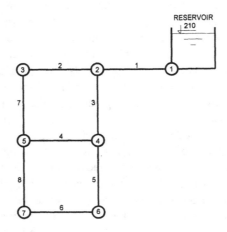

Figure 6.3.1 The two-loop network used as an example (by permission of ASCE [SA97]).

The network consists of a single source with a 210m head and eight pipes arranged in two loops. The pipes are assumed to be 1km long with Hazen-Williams coefficients of 130. The demand and head requirements at each node are listed in Table 6.3.1. The fourteen commercially available pipe diameters and their cost per metre are given in Table 6.3.2.

Node	Demand (m³/h)	Ground level (m)	Minimum Acceptable Head above ground level (m)
1(source)	-1120.0	210.00	30
2	100.0	150.00	30
3	100.0	160.00	30
4	120.0	155.00	30
5	270.0	150.00	30
6	330.0	165.00	30
7	200.0	160.00	30

Table 6.3.1 Node Data [SA97].

Diameter (in.)	Cost (monetary units)
1	2
2	5
3	8
4	11
6	16
8	23
10	32
12	50
14	60
16	90
18	130
20	170
22	300
24	550

Table 6.3.2. Pipe diameter and cost data [SA97].

VARIABLE FORM

The form of the unknowns in this type of problem is such that the mapping described in Chapter 2 between integer and reals is not required. Figure 6.3.1 shows that there are eight decisions (the eight unknown pipe diameters) to be made about the network. Each one of these decision variables can take one of the fourteen discrete values listed in Table 6.3.2. The shortest binary string that can represent fourteen values is of length four. However $2^4 = 16$ not 14. This illustrates redundancy in the coding, with two of the possible strings not representing any available pipe diameter. The simplistic way around such redundancy is by assigning the two additional string possibilities to other (close) pipe diameters. This will mean that in the two cases the same diameter is represented by more than a single sub-string, whereas the other twelve diameters are represented by unique sub-strings. Such an approach can have stochastic implications, although these are unlikely to be major unless the approach is taken to extremes.

GRAY CODING

The algorithm is based on a Gray binary coding of the decision variables. Such a coding represents adjacent integers in such a manner that they differ in only one bit position (see Chapter 4). Thus similar physical pipe diameters have similar string representations. In many problems this *adjacency property* is more important than one might initially think. For example, if the possible pipes in Table 6.3.2 are represented by a four-bit simple binary string, then the following ordered list is generated:

pipe size	4-bit binary representation	pipe diameter (in.)
0	0000	1
1	0001	2
2	0010	3
3	0011	4
4	0100	6
5	0101	8
6	0110	10

Moving from a pipe size 3 (4 inches) to size 4 (the next size up) cannot be achieved by flipping the value of a single bit, but requires three out of the four bits in the string to change their value, whereas in the Gray coding shown below, a single bit move gives a single increment in pipe size.

pipe size	4-bit Gray representation	pipe diameter (in.)
0	0000	1
1	0001	2
2	0011	3
3	0010	4
4	0110	6
5	0111	8
6	0101	10

Although by using such a coding, programming the algorithm is made slightly more difficult, the performance of the algorithm is likely to be improved. In particular, in the later stages of a run where the solution is near optimum and mutation is playing an important role, progression will hopefully become faster. Another advantage is that a particularly simple bit-based hill-climbing routine can be included as a direct search algorithm after the final generation has been processed.

THE PENALTY FUNCTION
The objective function, Ω, is particularly simple, just the sum of the cost of the individual pipes. However, each network created must be checked against the

minimum head requirements at the various nodes. The network solver used by Walters and Savic is based on the EPANET [ROS93] computer program. This uses a gradient method [TO87] for solving (6.3.2) and (6.3.3)

Unfeasible solutions, i.e. those which fail to meet (6.3.4) are not removed from the population. Instead their fitness is degraded. If they are far from feasible, they are likely to be viewed as "lethals" by the algorithm and fail to reproduce at next generation. Thus (6.3.1) is replaced by,

$$\Omega = \sum_{i=1}^{N_p} c(\Phi_i, \lambda_i) + \left\{ \max_j \left[\max\left(H_j^{\min} - H_j, 0 \right) \right] \right\} p$$

where p is a penalty multiplier and the term in braces equals the maximum violation of the pressure constraint.

The multiplier is chosen to normalise nominal values of penalties to the same scale as the basic cost of the network. Unfeasible solutions are likely to carry more information useful to the GA early on in the search than towards the end, when fine adjustments are being made to the optimal solution. In addition, in many search spaces there are likely to be more unfeasible solutions during the early generations. It would therefore seem sensible to devise a penalty term which becomes increasingly severe with generation. One possibility is:

$$p = \left(\frac{g}{G} \right)^{0.8} K$$

where K is a constant. When $g = G$, i.e. the final generation, p should be such that no unfeasible solution can be better than any feasible solution in the population.

REPRODUCTION

In an attempt to avoid premature convergence and the need for fitness scaling, rank selection was used with a linear weighting function.

After experimentation with different operators, uniform crossover was adopted. This also allowed (6.3.1), which describes a minimisation problem, to be used directly by simply ranking in reverse order. With $P_c = 1.0$, P_m was set to 0.03, or approximately $1/L$, and elitism applied.

AN ILLUSTRATIVE EXAMPLE

Although the network shown in Figure 6.3.1 is of modest scale, it still contains 14^8, or approximately 1.5×10^9, possible designs. This has forced all previous studies to consider split-pipe solutions. Table 6.3.3 lists some typical results, together with results from GANET. As different authors used differing values for the head-loss coefficients of the Hazen-Williams formula, the results are difficult to compare. Two GA-based solutions are presented covering the full range of published values. The results are particularly promising because of the use of realistic non-split pipes and because the calculations only took approximately ten minutes on a personal computer.

Figure 6.3.2 shows a typical plot of the cost of the best solution as a function of generation. The form is typical: a rapid reduction followed by much slower progress until the termination condition ($G = 500$) was reached. The best solution found was identified after approximately 130 generations. A population size of 50 was used, implying 250,000 evaluations were performed. At most, this represents less than 0.017% of the possible pipe combinations and therefore of the search space.

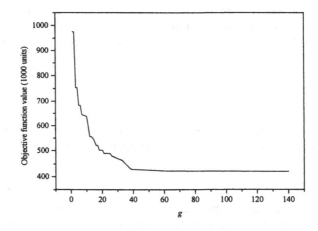

Figure 6.3.2. Evolution of cost with generation (data from [SA97]).

The quality of the solutions identified, and therefore the robustness of the method, can be gauged by hydraulic simulations of the other results presented in Table 6.3.3. These simulations were carried out for both the maximum and minimum values of the conversion constant ω in the Hazen-

Williams formula used to produce GA results. Table 6.3.3 shows that many of the results become unfeasible for at least one of these bounds. This is less so of the GA-based calculations. This indicates that the GA has identified robust engineering solutions to the problem.

Node i	Alperovits and Shamir [AL77]	Goulter et al. [GOU86]	Kessler and Shamir [KE89]	Eiger et al. [EI94]	GA No. 1 $\omega = 10.508$	GA No. 2 $\omega = 10.903$
2	53.80	54.15	53.09	53.09	53.09	55.86
3	31.89	32.78	29.59^{IF}	29.81^{IF}	29.97^{IF}	30.30
4	44.71	43.91	43.35	43.59	43.18	46.38
5	31.65	31.65	29.38^{IF}	29.90^{IF}	33.13	31.61
6	30.83	30.83	29.69^{IF}	29.46^{IF}	30.11	30.50
7	31.11	31.11	29.59^{IF}	29.34^{IF}	30.13	30.52

Table 6.3.3. Pressure heads for $\omega = 10.9031$, $a = 1/0.45$ and $b = 2.63/0.45$, (^{IF}, unfeasible solution: $H_i < 30m$) (data from [SA97]).

REAL NETWORKS

GANET has been applied to the trunk network of both Hanoi [FUJ90] and New York [SC69]. The Hanoi network consists of 32 nodes and 34 pipes organised in three loops, with approximately 2.9×10^{26} possible designs.

In the case of New York the problem was one of expansion, the goal being to identify the most economically effective designs for additions to the pre-existing system of tunnels.

In both cases the use of a GA provided robust solutions in a very efficient manner.

6.4 GROUND-STATE ENERGY OF THE ±J SPIN GLASS

Spin glasses have been studied for many years by physicists as idealised representations of the solid state. Such a representation consists of a large number of spin sites residing on an interconnected lattice (see Figure 6.4.1). However, interest in such systems goes beyond the world of physics and they are currently of great interest across a wide range of fields [BI86]. Additionally, computational methods developed to study spin glasses have been applied to questions in computer science, neurology and the theory of evolution [ST89,FA94,KI87]. A well known example of this is the design of artificial neural networks, where the Hopfield net is found to be isomorphic to the Ising spin model [RO93].

One problem studied within spin glass systems is the determination of the ground state energy for a system of infinite size. The value depends on the layout of the lattice, its dimension and the configuration of the bonds. Wanschura, Migowsky and Coley [WA96] have used a genetic algorithm together with a local search heuristic (as suggested by Sutton [SU94]) to estimate the ground-state energy of spin glasses of dimension greater than three. For a lattice of even modest size the determination represents a severe computational challenge. The work described here illustrates the following:

- long genotypes, $L > 3000$;
- large number of Boolean unknowns, $M > 3000$;
- inclusion of an additional local search heuristic (directed mutation), and
- direct correspondence between the problem-space and the string representation, negating the need to encode the unknowns.

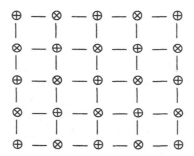

Figure 6.4.1. A regular spin glass in a minimum energy state showing spin sites ($\oplus$ = +, 1 or "up", $\otimes$ = –, 0 or "down") and bonds (| and —).

INTRODUCTION

The model used assumes that interaction only takes place between nearest neighbours. The number of such neighbours will however depend on the dimension D of the system ($D = 2$ in Figure 6.4.1). The interaction between any two spins i,k, is described by the coupling constant J_{ik} which represents the bond. In general, J_{ik} can take any value; however, here the restriction is that $J = \pm 1$.

Mentally, the problem is a simple one: given a particular configuration of bonds, find the configuration of spins that give the lowest energy (defined below) to the system. However, unless all the bonds are given the same value (for example the Ising model), the problem proves far from trivial. In particular, the following provide difficulties:

1. Frustration: In a typical spin glass system there is usually no spin configuration that will simultaneously satisfy *all* of the bonds: a local minimisation of the energy does not necessarily lead to the global minimum [TO77,VA77].

2. High non-linearity: The energy is a highly nonlinear function of the spin configuration, which results in many local energy minima.

3. Large configuration space: A system of dimension D and lattice length λ has 2^{λ^D} possible configurations and a large number of unknowns to be found.

These factors imply that an analytical solution is almost impossible, although mean-field approximations have been proposed. Thus, finding the ground-state of a spin glass has become a prime testing ground for numerical optimisation methods. Different methods have been proposed and tested (e.g. simulated annealing [DA87,RO93]).

METHOD
The very large number of possible combinations of spin orientations (2^{1000} for a system of dimension 3 and length or side 10) has lent to a certain reluctance [ST94] amongst workers in the field to tackle systems of higher dimension using traditional search methods. Recently it has been suggested [SU94] that a GA might provide a method for estimating the ground-state energy of spin glass systems of dimension greater than 3. The scale of the problem, combined with the existence of multiple local optima (at all scales), makes higher-dimensional spin glasses ideal subject matter for the testing of the new heuristic and connectionist search models now being applied within the physical sciences.

The ground state energy is defined as the minimum value of

$$H = -\frac{1}{2}\sum_{ij} J_{ij} S_i S_j \ , \tag{6.4.1}$$

for a system of spins S_i. The bonds J_{ij} are regularly oriented in the Ising model and randomly oriented in the Edwards-Anderson model [ED75,SH75]. The summation is carried out over nearest neighbour pairs. With randomly oriented bonds it is not possible to minimise (6.4.1) at each individual spin site and the value of the ground-state energy will depend on the particular bond configuration in question.

For the Ising model the ground-state energy E_{min} has a known value (per spin) for systems of any dimension D, and any linear size (or length) λ, and is equal to $-D$ (in units of J). For randomly oriented bonds it is believed that E_{min} (for infinite λ) has a value of -1.403 ± 0.005 for a 2-dimensional lattice [SU94,SA83] and -1.76 for 3-dimensions [SU94,CE93].

In the case of a spin glass, the encoding of the unknown parameters (S_i) into the genotype is particularly natural. Each individual in the population is a single complete lattice with 1 being used to represent an up spin and 0 to represent a down spin. The concatenated string is then just an ordered binary list. The configuration of the bonds is randomly assigned at initialisation. The formation of the next population is simply a matter of cutting pairs of genotypes at the same, random, point and swapping the severed halves of the lattices (i.e. single point crossover). Mutation is implemented by the occasional flipping of digits in the new strings. Selection of individuals to undergo this processes is inversely proportional to each individual's value of H, i.e.:

$$f = \frac{1}{H} \ .$$

For small lattices (and no mutation operator), convergence can be defined as the point where all strings in the population, and therefore all lattices, are identical. For large lattices such a strict definition of convergence can lead to long run times. In this work convergence is deemed to have occurred when at least 40% of the lattices are identical and the best estimation of E_{min} has not changed for 100 generations. It was found that convergence could be reached sooner if each spin-site in each new lattice was visited "typewriter", fashion with the spin being flipped if this led to a decrease in the energy of the *local* system (this is similar, but not identical, to the method used in [SU94]).

The algorithm was checked by reproducing the most recent estimates [SU94] of E_{min} for the Ising model and for the two- and three-dimensional Edwards-Anderson model cited above. This also allowed appropriate values to be found for the population size, the crossover rate and the mutation rate to ensure rapid convergence. A mutation rate of 0.001 was used together with a population size of 800 simultaneous spin systems, with 70% of the lattices undergoing crossover each generation (i.e. $P_c = 0.7$).

The GA was run with dimension two to six with L running between five and twenty and periodic boundary conditions (i.e. any side of the lattice in Figure 6.4.1 is joined to its opposite). Ten runs of the program, and hence ten different bond configurations, were tried in all cases except the case of $D = 6$ when only five runs were completed. The results are shown in Figure 6.4.2. Extrapolation of the results (using linear regression) to infinitely sized systems allows the estimations of E_{min} ($\lambda \rightarrow \infty$) given in Table 6.4.1.

Applying linear regression, once more, to the results given in Table 6.4.1 allows the decrease (in units of J) in E_{min} ($\lambda \rightarrow \infty$) with increasing dimension to be estimated 0.312 ± 0.0116 per extra dimension.

Dimension, D	E_{min} (in units of J)
2	-1.404 ± 0.005
3	-1.763 ± 0.002
4	-2.054 ± 0.003
5	-2.347 ± 0.016
6	-2.553 ± 0.022

Table 6.4.1. Estimated ground state energies, E_{min}, for infinite systems of varying dimension.

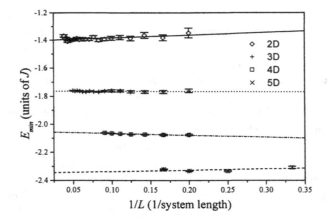

Figure 6.4.2. The calculated ground state energies, together with their standard errors, of the different systems (lines estimated by linear regression), from [WA96].

6.5 ESTIMATION OF THE OPTICAL PARAMETERS OF LIQUID CRYSTALS

Liquid crystals are finding an increasing number of applications from watches to complex displays for portable computers. This utility arises from the ease and speed with which the molecules within liquid crystals can be made to reorientate themselves. Interestingly, the precise orientation as a function of depth is not fully understood for many systems.

The best methods currently used to study this orientation [YA93,YA93a] rely on laborious hand-fitting of experimental data to a model of the system. Recent attempts to automate this process using a GA [MIK97,MIK97a] have proved highly successful. In order for the approach to work, a series of adaptations to the simple GA are required, and the approach illustrates much of the wrestling often undergone to produce a successful application. In particular, the following are considered:

- the development of a system to replace a previously human-guided search;
- how, starting from a simple problem, a series of sequential adaptations are made;

- the use of linear fitness scaling, bound reduction, fitness function adaptations, approximate fitness evaluation and a variable mutation rate; and

- the inclusion of a traditional search routine to create a hybrid algorithm.

LIQUID CRYSTALS

Liquid crystals exist in a series of states of matter somewhere between the usual solid and liquid states. They have been investigated since the 1880's when Reinitzer and Lehmann first studied esters (benzoate) of cholesterol. Unlike the usual states of matter, they display a series of "mesophases", each of which has distinct structural properties categorised by the degree of positional and orientational ordering found within them. It is the existence of these mesophases, and the possibility of making transitions between them, that have allowed liquid crystals to be deployed in a range of technologies.

The characteristic feature which links the many thousands of substances, both naturally occurring and manmade, that show liquid crystalline phases is that the molecules are anisotropic: either elongated, cigar-shaped or disk-like. The anisotropy of the molecules is reflected in intermolecular forces. It is these anisotropic forces which lead to the formation of the numerous mesophases. The average direction of the principal molecular axis is given by a unit vector, or *director*. As the temperature of a liquid crystal is reduced increasing order is gained. A typical sequence would be:

isotropic, which has no long range positional or orientational order (i.e. liquid like);
↓
nematic, which has no positional order ,but some orientational order;
↓
smectic, which has both positional and orientational order; and
↓
crystal, which has high positional and orientational order.

Many phases are chiral, i.e. the director precesses as you move through the material.

Not only are such molecules physically anisotropic, they are also dielectrically and optically anisotropic. As a result, if an electric field is applied alignment of the molecules will occur. This potential for alignment is at the heart of such materials being useful in display technologies. The optical anisotropy results in differing effective refractive indices for polarisation along, and perpendicular to, the molecular axis.

Sandwiching the liquid crystal as a thin layer between two glass plates that have rubbed polyamide coating their inner surfaces is enough to physically align the liquid crystal along the rubbed direction. If the glass plates are first coated with a transparent conductor then, once assembled, an electric field can be applied to re-orient the liquid crystal. Furthermore, if this cell is placed between crossed polarisers the alignment of the liquid crystal will effect the transmissions of light through the cell, resulting in a simple display system.

In a typical, standard, simple twisted nematic cell, about 10μm of liquid crystal is placed between the plates, with the alignment layers used to impose a 90° twist on the director through the cell. Incident light is then naturally guided (twisted) as it passes through the cell and is able to exit through the second polariser. Applying an electric field (of the order of 10^6 Vm^{-1}) induces dipoles in the liquid crystal molecules which align in the direction of the field, stopping the light from emerging from the second polariser. Removing the field returns the cell to its original state.

If a liquid crystal cell of low order is mechanically deformed, birefringence colours become visible, but once the deforming force is removed the molecules flow back to regain their original structure. However, in highly ordered systems in more solid-like mesophases, the structure of the system is left permanently deformed, making the device useless as a display. If more were known about the details of the exact structure within such cells (with and without defects) progress might then be made on better cell designs. Work by Mikulin, Coley and Sambles using GAs [MIK97,MIK97a] has been centred on detailing this structure.

THE HALF-LEAKY GUIDED MODE TECHNIQUE

The principal optical tool for the study of liquid crystals is that of optical polarising microscopy. Because this technique integrates the optical response through the entire thickness of the cell, only some weighted average of the director orientation is obtained. In order to understand the underlying structure within the cell, a method that can detect details of the spatial variation of the director is required. The development of optical waveguide techniques, in particular the use of the Half-Leaky Guided Mode (HLGM) technique by Yang and Sambles [YA93,YA93a], provides just such a method.

A waveguide consists of a sandwich construction, such as a fibre optic, where a core of high-index glass is surrounded by glass of lower index. The difference in the refractive indices can ensure, under the right circumstances, that the light is contained within the waveguide. The light in such a waveguide *mode* follows a zigzag path within the central medium. Many such paths or modes are possible. In many ways these modes are similar to organ-pipe standing waves. The lowest energy mode that can be excited is the fundamental, which has averaged electric field distribution nodes near the waveguide surfaces and a maximum in the centre. Higher energy modes are harmonics where progressively more half-wavelengths fit into the waveguide. The maxima for each of these harmonics are seen to occur at different positions within the waveguide. By building a waveguide consisting of a liquid crystal layer, the optical excitation of the guided modes which may propagate in the cell may be used to characterise the liquid crystal optical parameters as a function of position within the cell. In particular, measuring the angular dependence of reflectivities which characterise resonant modes provides data, which when compared to a mathematical model of the cell gives the director profile through the cell.

In practice, polarised monochromatic laser light is incident (via a prism) upon the cell which sits on a rotating table. A photodiode detector is used to detect reflected light, and an oven enclosure is provided to allow the study of various mesophases. The incident beam can be either p- or s-polarised and the detector is arranged such that only p- or s-polarised light is detected. This means that four possible angle-dependent datasets of reflected light can be collected: R_{pp} (p-in, p-out); R_{ss} (s-in, s-out); R_{ps} (p-in, s-out) and R_{sp} (s-in, p-out). A typical data set contains measurements made at about 1,000 different angles.

THE SEARCH SPACE

The mathematical model used of the cell is based on Fresnel's equations of reflection, and views the cell as a series of discrete optical layers. These layers correspond either to real physical layers in the non-liquid crystal part of the cell, or arbitrary sub-layers of the liquid crystal chosen such that any individual layer is optically thin. The number of sub-layers required to accurately represent the liquid crystal depends, in part, on the complexity of the liquid crystal mesophase in question. As the number of sub-layers grows, the numbers of unknown adjustable parameters (used to describe how the molecules bend and twist the light) in the model also grows. Attempts to find a traditional search method capable of adjusting these parameters, so as to minimise the difference between the experimental data and predictions from the model have failed because of the complexity of the search space. This has meant that such fitting is typically carried out by a human-guided search through the space with intermittent use of a gradient-based search algorithm. Such a search may take several man-months.

The complexity of the search space arises from the large number of parameters (typically more than 50) which are being adjusted and the existence of a very large number of local optima throughout the space. This results in many combinations of parameters leading to similar traces. One way around this problem is to attempt to fit more than one dataset, i.e. R_{pp} and R_{ss} at the same time. This removes many of the degenerate solutions because some often prove much poorer solutions for the second dataset. Another possibility is to change the azimuthal orientation of the cell relative to the polarisation of the incident laser beam and collect new data sets. Given enough data it should be possible both to navigate through the space more effectively and also locate a definite global optimum. However there are problems with such an approach: the more data the better the fit, but the bigger the search. This is a general problem with many experimental datasets. Collecting more data makes the topology of the modelled search space closer to the physical problem space, smoothing out many irregularities and, hopefully, allowing the identification of a global optimum, but only at the cost of a much larger problem.

In the HLGM data, if there are R_{pp}, R_{ss}, R_{ps} and R_{sp} sets at 1000 angles and two azimuthal angles, then the computer model will have to make 4 x 1000 x 2, or 8000 estimations of the reflectivity for a single guess at the unknown parameters, making it impossible to consider random or enumerative searches and difficult for any other method. These problems—of a complex search space together with computational time constraints—have meant that it is considered impossible to successfully fit such datasets, and thus impossible to fully characterise such cells.

The use of a GA on the problem proved highly effective, but not without difficulty. The first GA applied was very simple indeed. A binary encoding was used (with 10 bits per unknown, i.e. $l_i = l_j$, $i \neq j$) together with rank order selection (with the best 50% of individuals being selected for crossover once and only once) and random replacement by children of the original population members to create the new generation. Mutation was applied, at a rate of 0.001 per bit per generation, across the complete new generation, and elitism applied. The fitness measure used to distinguish between solutions was based on the sum of squares (SOS) difference between the experimental and modelled data:

$$fitness' = A - SOS$$

where

$$SOS = \sum_{i=1}^{k} (R(\theta_i) - \rho(\theta_i))^2 \, ,$$

where R is one of the reflectivities R_{pp}, R_{ss}, R_{ps} or R_{sp} and ρ is the theoretically predicted (i.e. modelled) reflectivity. The sum is carried out over all data points, or angles, θ. The constant A is selected to be equal to or slightly greater than any likely value of SOS. On the odd occasion when a negative fitness was produced, the fitness was set equal to zero. The results proved surprisingly good for a first attempt. Typical initial values of SOS of 200 were reduced to 0.23 within a few hours on a fast personal computer for a nine parameter nematic (i.e. relatively simple) cell problem and gave parameter values very close to those found by laborious hand-fitting (an approach which might take a month for such data if similar bounds were used).

Initially, attempts to study more complex cells with 27 unknowns failed. When presented with both R_{pp} and R_{ps} data at two azimuthal angles, where the minimum value of SOS was known to be approximately 10^{-3}, only SOS's of something less than 10 were achieved. In order to make the approach useful for characterising more complex cells, a series of adaptations were made to the algorithm.

REDUCING THE STRING LENGTH

The first improvement tried was to assign different binary word lengths to different parameters. This very basic adaptation was felt to be sensible because different parameters were known to affect the goodness of fit to varying degrees. This is probably crucial for a GA to work effectively with a large number of unknowns. A simple GA with a string length of several thousand is likely to take a very long time to converge within most real search spaces.

Such an approach rapidly reduced (in < 100 generations) the SOS to less than 1. However, progress past this point proved very slow. One reason appeared to be that for $SOS < 1$, many of the strings have bit-patterns very close to the pattern they would have at the global minimum. At such a point it becomes very difficult for a GA (with large L) to progress by simple crossover and mutation to the global optima. In particular, mutation will almost always be destructive and the search becomes close to looking for a needle in a haystack. Unfortunately switching to a traditional search mechanism at this point could still not guarantee locating the global optimum. One approach to this problem proved to be the sequential use of the GA within a narrowing search space. This ensures an efficient gridding of the search space by the binary encoding. A sketch of the method is given in Algorithm 6.5.1.

> 1. Apply GA until the solution no longer improves or for a fixed number of generations.
> 2. Reduce the size of the search space and re-initialise the GA
> 3. Repeat from 1 a set number of times, or until convergence is reached.

Algorithm 6.5.1. A GA with bound reduction. Note that such an approach is not suitable for most types of search space and care has to be taken not to exclude regions of the search space too early.

Step 2 of Algorithm 6.5.1, the reduction of the size of the search space, could take one of many forms. The simplest would be to reduce the width of the space for each unknown by the same fractional amount, giving a new space centred upon the current best solution. This could well cause problems. For example, if the old space had width w_j, for parameter j and the new space is to have width $w'_j = w_j/2$, then if the current best guess of parameter j is closer than $w_j/4$ to the upper or lower bound of the original space, then the new space will include a fragment outside of the range initially specified as the problem space.

At best the inclusion of such a fragment is wasteful, at worst it may lead to the production of an unfeasible solution. One way around this is not to allow any new bound to fall outside the initial search space, and if it attempts to do so, to simply give it the value of the initial bound.

Another question to be answered is the size of the reduction constant n_j, where:

$$n_j = \frac{w_j}{w'_j} .$$

The greater the value of n, the faster, but the less reliable, the convergence. One possibility is to make n_j dynamic and base its value on the value of σ_j (or some other statistic) estimated from a single run or a series of runs of the GA. This could occasionally allow $n_j < 1$ (i.e. an expansion of the space). If all runs progress to similar values of a parameter then a greater reduction in the size of the search space is possibly justifiable. However, extreme caution is required to make sure that at no point is the true global minimum excluded from the search (this will not be possible for many search spaces).

Using such an approach on the HLGM data proved very successful, with the *SOS* rapidly being reduced toward zero (Figure 6.5.1).

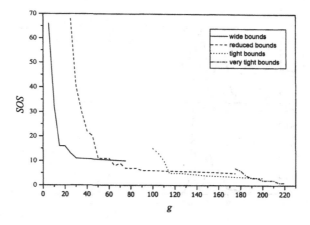

Figure 6.5.1. Enhancing the GAs performance by repeated bound reduction (data from [MIK97a]).

In essence, such bound reduction is attempting to make the crossover and mutation operators equally effective across all unknowns, regardless of the sensitivity of the problem to a particular parameter. Hence for parameters which initially have little effect on the fitness, the bounds will remain wide; for the more critical ones the bounds will slowly reduce. There are many functions for which such an approach is unlikely to be suitable and will flounder. However, within spaces characteristic of several physical problems where a solution can be incrementally *zoomed* in upon it may be successful. An example of the successful performance of the method on *theoretically* produced HLGM data for a liquid crystal in the nematic phase is shown in Figure 6.5.2.

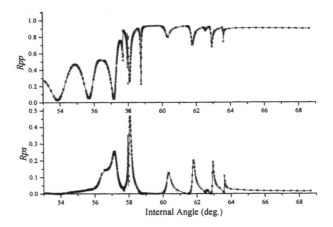

Figure 6.5.2. The performance of the sequential GA on the relatively simple problem of nematic HLGM data with 27 unknowns: cross = data, line = fit) (data from [MIK97a]).

APPROXIMATE FITNESS FUNCTION EVALUATION

As has been suggested before, the time it takes for a GA-based optimisation routine to cycle through a generation is likely to be dominated by estimations of the fitness function, rather than the genetic operators. In the case of the HLGM data this means that the computing time required is proportional to the number of data points. However the accuracy of the final answer will, in some complex way, be a function of the number of data points. So a conflict exists between speed and accuracy. Or, put another way, given a fixed (or reasonable) amount of time for the running of the program, how can accuracy be maximised? Is it best to estimate the fitness function as accurately as possible and reduce the number of generations processed, or would it be better to use an approximation to the fitness function (which can be estimated more rapidly) and process more generations? The answer to this question is obviously problem-dependent. It is also likely to change during the optimisation process itself. During the early stages of the search, fitnesses are likely to be low and so their accurate estimation is unlikely to be important. However, at later stages navigation of the algorithm through the search space is likely to rely on subtle differences in the fitness function in different directions of the hyperspace.

To try to speed up the search, and thereby ultimately allow for greater accuracy, the HLGM data was filtered to only leave $D\%$ of the data points. As the run progressed the value of D was increased. For the HLGM data the form $D = D(g)$ was used; however $D = D(f)$ would possibly be more natural. With this approach, it was found possible to quarter run times for no loss of accuracy.

USE OF FITNESS SCALING

As discussed in Chapter 3, GAs can experience problems if the fitness function spans too little, or too great, a range within any one generation. In particular, if during the early stages of a run a small sub-population (typically of size ≈ 1) has individuals with a much greater fitness than the population average then the sub-population can rapidly grow under fitness-proportional selection to dominate. This can lead to premature convergence. Conversely, during the later stages, a large sub-population (typically of size $\approx N$) may only contain individuals who have fitnesses approximately equal to the population maximum. In such a situation, simple fitness-proportional selection can make little distinction between sub-population members and the progression of the algorithm is much reduced.

Linear fitness scaling was used on the HLGM problem to control the expected number of times above-average population members would undergo selection, compared to an individual of average fitness. A range of values for the multiplier c_m was tried and an improvement in convergence noted. Making c_m dynamic, i.e. $c_m = c_m(f)$ or $c_m = c_m(g)$ was not tried.

DIRECT FITNESS FUNCTION ADAPTATION

Unlike many problems where few details of the functional form of the search space are known prior to optimisation, partial enumerative searches through theoretical data had produced insights for some cells. In particular, many of the unknowns had been found to show sections through the solution vector that contained large areas of equal fitness near f^*. Such a section suggests the substitution:

$$f' = f^m \; ; m > 1 \, ,$$

might be of benefit. Such direct scaling can only be applied if some of the details of f are already known (allowing a sensible choice of m).

For the HLGM problem several values of m ($m = 6$ proving the most useful) were tried and again improvement in convergence witnessed across many sets of data. (Similar results could have been realised by making suitable adjustments to the selection mechanism itself, because fitness function adaptation and fitness scaling are two sides of the same coin. However, it is often more naturally intuitive to adapt the fitness function directly.)

BOILING THE POPULATION

Long periods of only marginal, or no, progress were seen during many runs with some cells. An attempt to see if the computer time during such periods could be better spent was made by making the substitution

$$P_m' = BP_m \; ; B > 1 \; , 0 < P_m \ll 1 \; , 0 \ll P_m' < 1 \; .$$

In essence the population was *boiled*, whilst maintaining elitism, by momentarily (typically for three generations) increasing the mutation rate. It was hoped that not only would this encourage diversity within the population, but also allow the population to jump over traps in the binary representation caused, in part, by not implementing a Gray coding.

In order for such an approach to effect the diversity of the population for more than a few generations after the boiling event, fitness scaling must be applied simultaneously. Without such scaling most individuals produced by the operation will be lethals (except the elite member) and fail to be selected for subsequent generations. This reinforces the idea that additional GA operators cannot be used in isolation and without regard to those operators (and their settings) already in use by the algorithm.

ADDITIONAL DIRECT SEARCH

In line with the comments made in Chapter 1, many problems are likely to benefit from the inclusion of a traditional search algorithm working in combination with the GA. In this case this was done by the inclusion of a direct search method proposed by Jeeves and Hooke [BU84], although other methods could have equally been used. The direct search was applied (to data from the highly complex smectic C* phase with 55 unknowns) after $g = G$ in an attempt to climb the final hill. This was successful, with the improvement in fitness being far in excess of that which might have been expected from a similar number of additional function evaluations within the GA.

The final (and most successful) algorithm used in this work can be described by the schematic given in Algorithm 6.5.2. The algorithm was

arrived at by trial and error based on an extensive knowledge of the macro-
details of the search space but has little justification other than it works
extremely well for the problem at hand. If such knowledge had not been
available, then some of the settings within the algorithm might have been
discovered by including them in the search space, using the techniques
described in references [GR86, BR91, DA89 and DA91].

---

1. Run GA with $G = 30$, $n = 60$, $P_c = 0.5$, $P_m = 0.0016$, linear fitness scaling,
   fitness-proportional selection with elitism, and using the minimum substring lengths required to maintain accuracy.
2. Boil population, maintaining the elite member.
3. Repeat steps 1 and 2 six times.
4. Reduce the bounds on the parameters to 120% of the spread of values of
   f_{max} $(G = g)$ obtained from the six GA runs.
5. Repeat from 1 five times.
6. Apply direct search.

Algorithm 6.5.2. The final algorithm used with the HLGM data.

This algorithm is a long way from LGA and indicates that GAs can
benefit greatly from adaptations that attempt to include additional knowledge
about the search space in question. The success of the method can be gleaned
by the closeness of fit shown in Figure 6.5.3, which shows a fit to the much
more complex smectic C* phase. Figure 6.5.4 shows the liquid crystal
molecular orientations discovered by the GA: the molecules are seen to
gradually twist and tilt through the depth of the crystal.

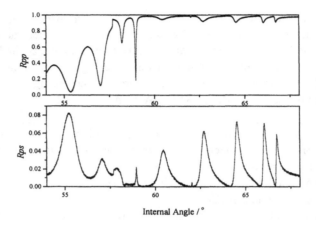

Figure 6.5.3. Fit (line) to experimental smectic C* data (at 21°C) (+) with 55 unknowns (data from [MIK98]).

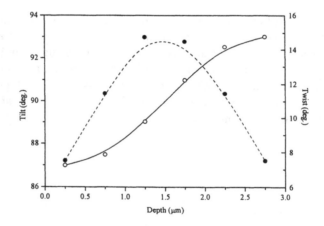

Figure 6.5.4. The twisting (●) and tilting (○) of the molecules through the depth of the crystal discovered by the GA (points connected by B-splines) [MIK98].

6.6 DESIGN OF ENERGY-EFFICIENT BUILDINGS

Considerable scope exists for reducing the energy consumption of both new and existing buildings.

Although some reductions in energy use can be achieved by relatively simple measures, very high levels of performance require the coherent application of measures which together optimise the performance of the complete building system. In essence, the designer, who will typically be more of an architect than a scientist or engineer, is faced with a multi-component optimisation problem. Most designers feel ill-equipped to tackle such a task and this is a serious obstacle to them advancing high performance designs. The application of computerised optimisation techniques to the design of low energy buildings could provide architects with a powerful new design tool. With the increasing use of graphical packages during other stages of the design progress such an approach could eventually be included within the design environment.

Population-based optimisation appears to be ideally suited to providing the type of support and assistance needed. Most traditional optimisation techniques that might be applied suffer from the drawback that only a single result is obtained. Because of the difficulty of including such factors as aesthetics in the optimisation process (as discussed in Chapter 4), it is likely that any "optimum" result will be found to be unacceptable.

Given the number of individual attributes that combine to make a single building, the number of possible realisations, or designs, is very large. Work by Migowsky [MIG95] showed that a GA can be used to allow a rapid and efficient searching of this multi-dimensional building space. The method produces not only the near-global optimum, but also a *set* of high quality designs. Each has excellent energy performance, but the set is sufficiently diverse in physical characteristics to allow the designer the opportunity to select on the basis of other important non-optimised criteria. A GA can thus form the basis for a powerful and practical building-design aid.

The system allows a designer to select from a range of buildings, all of which have a predicted energy performance within, say, five percent of the achievable minimum, the design which best suits his/her other requirements. This approach is in line with the established preference of architects to work from case study and demonstration material. It relieves the designer of the need to work through the consequences of choosing individual features and of checking their compatibility with the rest of the building system. The architect is instead offered a selection of optimised, complete, building packages from which to make a selection.

The complete system has a pair of criteria to be satisfied: first a numerical measure of fitness which can be used to drive the GA, and second, a qualitative assessment of the aesthetics of the design to which it is impossible to attach a numerical estimate. Because this second measure is non-numerical, Pareto optimisation methods (see §4.4) are not suitable.

Although the energy model used in this work contains only five true unknowns, it is impossible to describe or visualise a building with only these parameters. The building must be considered as consisting of a far greater number of parameters (>100) which are far from linearly independent. In essence, the GA is being used to throw up ideas, good ideas, because they are highly energy efficient, but still only ideas.

In the work described here, which is taken from developments based on reference [MIG95], the approach is applied to the design of a set of school classrooms. The work shows an example of a problem with the following features:

- a range of different variable types (binary, integer and real);
- the need for a diverse range of approximate solutions together with the global optimum;
- the need for human-based final selection;
- a large multimodal space studied by using multiple runs;
- a highly non-linearly independent search space;
- because of the time taken to evaluate a single design, the need to avoid re-estimating the fitness of any previously processed design;
- remainder stochastic sampling to help reduce the convergence rate; and
- use of a generation gap.

INTRODUCTION

Up to fifty percent of the United Kingdom's consumption of fossil fuels is related to energy use in buildings. The burning of fossil fuels, and the associated emission of combustion products, contributes to global warming, acid rain and other local pollution.

The energy performance of a building is determined by its response as a **complete system** to the external environment and the internal environmental demands of the occupants. The system response, in turn, depends upon the combination of individual attributes that have been assembled to produce the building. Thus, for example, a building with large windows, lightweight structure, elongated form and a powerful heating plant may be more, or less, energy-successful in the role of a primary school than one with a different mix of features. Traditionally, a solution is proposed on the basis of experience and on the evidence of the performance of other demonstrations. The performance is then checked using predictive models, and the design may be adjusted in the light of the results. By this means, an acceptable solution is arrived at—but there is no practical way of determining how close to a realistic optimum the final design is.

Unfortunately for the designer, it is not possible *a priori* to say which detailed combination of attributes represents the best solution to a particular brief. This situation arises because, although the performance of any particular system can be predicted using a suitable mathematical model, the inverse problem of determining the optimum system characteristics from the desired performance cannot be solved. Optimisation is therefore a process of trial. Given the number of individual physical attributes comprising a single building, the number of possible combinations which results from varying each attribute over its range of practical values is enormous. This precludes direct modelling of the entire building set and indicates that the multi-dimensional building parameter space has to be searched in a more efficient manner.

The main characteristics of the problem are a large multi-dimensional space to be searched, requiring an efficient method to converge within a reasonable time; a multimodal space requiring techniques to avoid convergence on false minima; a requirement not just to identify a single optimum result, but to also map the positions of a range of lowest energy solutions throughout the space.

The method described in reference [MIG95] uses energy predictions for individual buildings made using the building thermal simulation EXCALIBUR [CR87a, CR87b]. However, any other proved thermal modelling approach (many of which are far more sophisticated) would be suitable.

THERMAL MODELLING
The five-parameter computer model of the thermal response of buildings developed at Exeter, EXCALIBUR, has been used to aid the design of over seventy large buildings, from schools to underground military facilities. The success and continuing utility of this model has confirmed the original conviction that the essential features of building performance, in terms of internal temperature evolution and total energy usage, can be adequately represented by a lumped-parameter dynamic model. This approach results in a

160

small portable program capable of running quickly on desktop machines. The model is based on a two time-constant analogue circuit with five adjustable parameters per heated zone of the building (Figure 6.6.1).

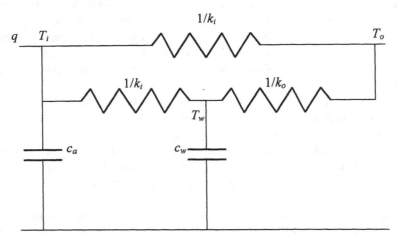

Figure 6.6.1. The electrical R-C analogue of a single EXCALIBUR zone.

The characteristic equations are:

$$c_a \frac{dT_i}{dt} = k_i(T_w - T_i) + k_l(T_o - T_i) + q$$

and

$$c_w \frac{dT_w}{dt} = k_i(T_i - T_w) + k_o(T_o - T_w)$$

where k_l represents a quick response thermal conductance (ventilation heat loss, window losses etc.) between the inside temperature node T_i and the outside temperature T_o; k_i is the thermal conductance between T_i and the mid-structural temperature T_w; k_o is the conductance between T_w and T_o; c_a is the effective thermal capacity of the air; c_w the thermal capacity of the structure of the building; and q the heat supply (including metabolic, lighting and other gains) acting at T_i.

These equations are solved analytically and the model building is time-stepped through a complete season of representative weather data, including angle dependent solar intensity. Realistic occupancy and temperature schedules can be specified and the user can choose between manual and optimum-start plant control. The program can accommodate up to ten thermally linked and separately controlled zones.

The model therefore provides the means whereby reliable predictions of seasonal energy use can be made for buildings with any reasonable combination of physical attributes. However, such a model suffers from at least one major setback: few architects (or building scientists) would be able to visualise designs in terms of the model's parameters, V^{mod} (i.e. c_w, k_o etc.). The designer is using the much larger set of physical attributes, V^{phys}, such as positions, wall lengths, material types, etc, from which the five model parameters per zone are estimated. Many combinations of physical attributes will give rise to near identical combinations of model parameters. Many attributes are highly related—the perimeter and the enclosed volume for example). The search space is therefore highly multimodal and chocablock with non-linearly independent variables. The scale of this problem can be illustrated by the realisation that the search space may contain in excess of one hundred unknowns, yet the model contains only five.

THE PROBLEM

The problem at hand is that of finding diverse sets of vectors V^{phys} such that the energy use of the building over a year is near minimum. The ideal value, T_{set}, of the internal temperature, T_i, during occupancy is set by the designer and the amount of energy required to maintain this value will depend greatly upon the design. The building gains "free" heat from solar input through windows, metabolic gains from the occupants and gains from electrical equipment (for example the lighting system). If these gains are insufficient to provide T_{set} then additional input is provided by the heating system. If $T_i > T_{set}$ heat is removed by the cooling system. In a low energy building, the need to provide such additional gains must be kept to a minimum. This position is achieved by "minimising" losses (k_o, k_i, k_l) and "maximising" the use of free gains. The thermal capacitance of the building fabric is then used to try to time-table the match between supply and demand. Unfortunately, this proves to be a fine balancing trick. If solar gains (for example) are truly maximised and losses simultaneously truly minimised, the building will frequently overheat, requiring the expenditure of large amounts of cooling energy.

The loss terms (k_o, k_i, k_l) and the thermal storage terms (c_w, c_a) are formed from a complex combination of the building components. This list of components is extensive. A single wall typically contains many layers: external brick, air gap, insulation, inner concrete-block layer, and plaster. Both inner and outer walls have to be considered, as do the floors and the roof. EXCALIBUR takes a single combination of these components as specified by the designer, forms a single set of model parameters (k_o, k_i, k_l, c_w, c_a), then combines these with an occupancy schedule and cycles the building through a year of weather data whilst using the heating and cooling systems to ensure $T_i = T_{set}$ during occupied hours.

In the work of Migowsky [MIG95], and subsequent extensions, EXCALIBUR is combined with a genetic algorithm to try and generate high-quality designs with architectural appeal.

The population, P, is manipulated as follows:

1. create initial, random, population of attributes $P(V_i^{phys})$; $i = 1 \ldots N$;
2. reduce physical attributes to model parameters $V_i^{phys} \rightarrow V_i^{mods}$; $i = 1 \ldots N$;
3. calculate annual energy use of heating and cooling systems, E_i ; $i = 1 \ldots N$;
4. use GA to create new designs V^{phys} by recombination and mutation using $f = A - E$ (where A is a positive valued constant) as the selection factor;
5. repeat from 2 until termination criterion is met; and then
6. filter designs considering architectural appeal.

It should be noted that human judgement is only applied after $g = G$, not, as in the example presented in §6.7, where such judgements are the driving force behind the GA's selection mechanism.

The success of the GA phase can be gauged from Figure 6.6.2. Here, a small extension to a primary school is being designed. The annual energy use of the building rapidly falls with generation, and the final design is seen to be far better than is typical for such structures.

REPRESENTATION

Each V^{phys} is represented as a binary string. As some parameters are reals (e.g. "building perimeter"), some are integers and others are Boolean (e.g. "double or single glazing"), the substring length l for each parameter varies greatly. This implies that the mutation and crossover operators interact with some parameters more frequently than others. If l is chosen to truthfully reflect the required accuracy of each parameter, this probably presents no problem because the degree of interaction is then proportional to the required accuracy. However if each l is not selected with care, then the algorithm will spend a considerable time processing irrelevant information. For example, if the wall thickness is only specified to the nearest cm, then there is little need to specify the building perimeter to the nearest mm. In reality, the correct set of sub-string lengths can only be identified given enough experience of the likely impact of each unknown on E, and knowledge of the commercially available sizes of constructional components.

POPULATION DIVERSITY

As the desire is to find an extensive range of possible designs, not just the optimum, population diversity must be maintained throughout the run. The algorithm used is a simple GA and therefore does not include niche and speciation methods; thus the selection and mutation operators must be relied upon to perform this role.

Selection is via remainder stochastic sampling (which places a limit on the number of times any individual can be selected) with a generation gap (see Chapter 4). Mutation is applied only to the children, which replace at random individuals in the original population. This approach is found to simultaneously maintain diversity and remove low quality designs rapidly. A population size of 50 and a generation gap of 0.8 are used.

AVOIDING DUPLICATE FITNESS EVALUATION

The evaluation of E_i for all i is time consuming, especially on the type of machine to which a building designer might be expected to have access. Thus it is crucial that designs are not re-evaluated. This is simple to achieve: a list of all unique designs, together with their respective value of E, is kept and new designs checked against this list. If a match is found then the corresponding value of E is simply assigned to the new individual.

This list-keeping process, or something similar, is probably worth including in many scientific or engineering-based applications of GAs. Only for the simplest of problems (typical of those used as test functions) is it likely that searching the list will take longer than a single fitness evaluation. By searching the list in reverse order (assuming new designs are placed in at the bottom) and not searching for any member which has not undergone crossover or mutation (because its fitness will not have changed), efficiency can be further improved. For long lists, it may prove desirable to only keep and search designs from relatively recent generations.

Although a reasonable level of genetic diversity is maintained during early stages of the run, Figure 6.6.2 shows that this diversity is not maintained throughout the run. Here the number of matches that were found between the current population and the ordered list is being plotted. (The algorithm never truly converges, which would be indicated by fifty (i.e. N) matches in any generation, because of the relatively high levels of mutation used.) It is thus clear that if this work is to be continued, additional measures will have to be taken to increase the degree of exploration in the algorithm during these later stages—possibly by never allowing two individuals within the current generation to be identical.

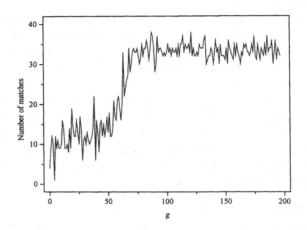

Figure 6.6.2. The number of matches (out of fifty) found between the current population and the ordered list of designs. Convergence (stagnation) is avoided by the use of relatively high mutation rates.

THE ADDITION OF ARCHITECTURAL APPEAL

The problem presented is a multicriteria one, requiring, at least, the minimisation of energy use and the maximisation of architectural appeal. No attempt has been made to include this appeal in the selection algorithm, and it is not clear—even using Pareto optimality—how this could be achieved without greatly reducing the number of processed designs.

As an alternative to even attempting this, Migowsky uses a filtered stacked histogram to present the best designs. The list of all designs (Figure 6.6.3) is initially filtered to remove all with a value of E more than $v\%$ greater than the best. (Typically $v = 5$ to 10%.) By using $\mathcal{A} \gg 1$ the regions around many local optima are included in this set (see §4.2). A stacked histogram of these unique designs is then presented on screen for the designer to study (see Figure 6.6.4 for a fragment). A select few of these designs can then be worked up to sketches (Figure 6.6.5).

The two designs shown in Figure 6.6.5 have relatively similar annual energy usage, but are of very different design. One uses high levels of insulation to reduce losses, the other maximises solar gains. Interestingly, combining these two design philosophies within a single building does not produce a high performing structure because it would overheat on sunny days (and thus have reduced fitness). The method is highly successful, not only as a method of obtaining high performing designs, but equally importantly, catalysing the design process by showing that similar environmental performance can be achieved in a wide variety of ways.

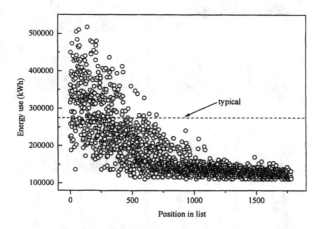

Figure 6.6.3. Results of designing a small low-energy extension to a primary school. The annual energy use of the designs is seen to reduce as the generations pass. The typical UK average energy use for such a building is shown for comparison [EE94].

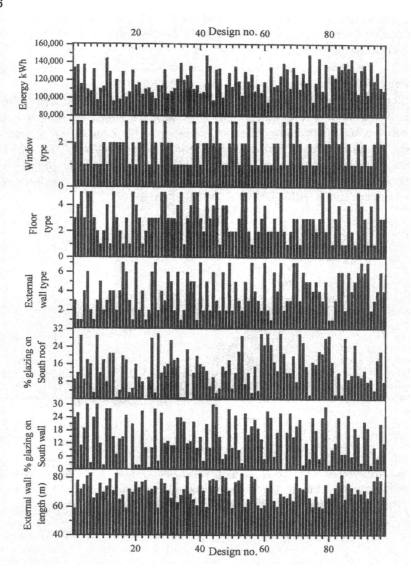

Figure 6.6.4. A fragment of the stacked histogram (taken from the data of Figure 6.6.3). The filtered designs are presented as a series of histograms, one histogram per building attribute. Individual designs can be identified by vertically connecting the histograms. The complete histogram contains many more attributes and all the designs in the filtered set.

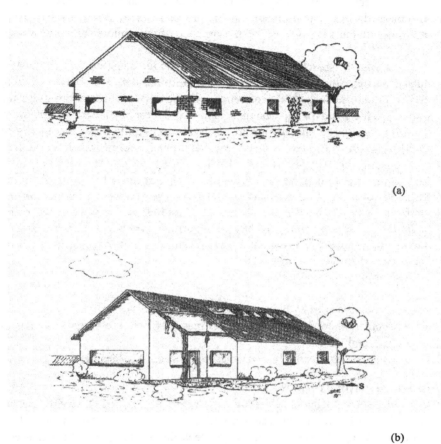

(a)

(b)

Figure 6.6.5. A pair of designs sketched from the later generations of a run [from MIG95]. The two designs have relatively similar annual energy usage, but are of very different design. (a) achieves high performance by minimising losses, (b) by maximising solar gains.

6.7 Human Judgement as the Fitness Function

Successful product design not only requires finding good engineering solutions to the problem at hand, but often also finding aesthetically pleasing solutions to customers' desires. The aesthetics of the solution can be extremely important regardless of the size, or cost, of the product—from children's toys to office blocks.

Aesthetic search spaces are typically large, complex and frequently shaped by individual, age, gender and cultural considerations [CR84,CO92b, SH83]. Often the designers' preferences will be very different from those for whom products are being designed. The work described here (taken from [CO97a]) represents one attempt to see how successfully a genetic algorithm could be in providing the designer with additional information. In particular, the question of whether a GA, when applied to objects encoded in a particularly simple manner (and driven by decisions made by a subject on the aesthetics of a potential product), would converge at a useful rate and in the direction of improving subject preference is addressed. Because of the many random processes operating within the algorithm, combined with the limited number of evaluations a human operator can be asked to perform, such an algorithm might well perform badly.

The problem demonstrates:

- a GA being driven by human judgement;
- a problem where identical genotypes may be given non-identical values of fitness; and
- a system where relatively few fitness evaluations are feasible.

Computer Aided Design

Computer aided design is becoming more common. The design environment is typically able to present the designer with a realistic view of the item and allow numerical analysis to be carried out to check the technical feasibility and cost of the design. The design environment may also allow optimisation calculations to be carried out. The optimisation itself will typically be for maximum reliability or minimum cost [AR89].

Any aesthetic optimisation routine within such an environment would have to contain a description of the aesthetic ideals of the target audience. This information is often gleaned from the preferences of a market-researched subgroup of this audience. A frequently-employed technique is to show a number of potential designs to a series of subjects and assess the impact of the designs upon the subjects. If these designs are actual mock-ups they will be very limited in number and there is the danger that one is only ascertaining the subjects' views of what the design team considers worthwhile; thus the subject, and hence the potential customer, is not being truly involved in the design process.

Computer aided design offers a way out of this trap. The computer can offer an almost unlimited number of designs to the subject as visualisations on the computer's screen. Such a design system can refuse to allow designs that are unsafe, unfeasible or too costly to implement within the production environment. However, the question remains of how the design tool and the subject should guide each other through what may be a large search space.

THE GA DESIGN STRATEGY

Several authors have attempted to use a GA with subjective evaluation, including work on dam design [FU93], facial recognition [CA91] and colour preference [HE94]. Coley and Winters [CO97a] decided to estimate the efficacy of the GA for searching an aesthetic, but realistic, commercial product space.

It was decided to choose a realistic product in which the aesthetic aspects of the design would be considered paramount, and one with few engineering constraints. The chosen product was a 1m square wall-hanging containing simple geometric designs. Shapes were ascribed one of 16 basic colours. Only isosceles triangles, rectangles, ellipses and straight lines were considered for these basic shapes. A maximum of 20 shapes (or objects) were present in any one design. The designs were processed by the GA as concatenated binary strings representing the positions of corners, lengths and the colours of the separate objects. Although simple, the designs were realistic and could be considered reminiscent of some of the designs contained within paintings by Carla Prina, Josef Albers, Piet Mondrian and Jean Gorin [SE62].

The optimisation process was very simple:

1. the computer generated an initial random population of 32 designs together with their string representations;
2. the subject viewed each design and scored each one for appeal;
3. the GA was used to produce new novel designs by crossover and mutation of the strings; and
4. the process was repeated from step 2 for a total of twenty generations.

Figures 6.7.1(a) to 6.7.1(f) (and the jacket cover) show a series of typical designs.

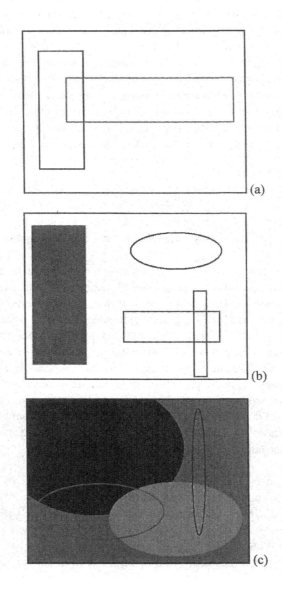

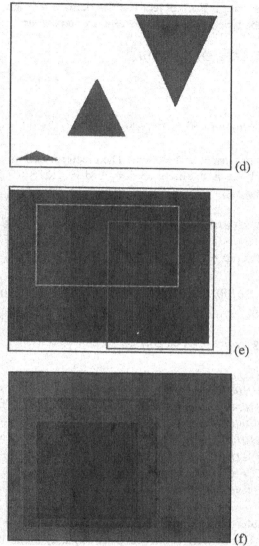

Figure 6.7.1(a-f). Black and white renditions of a selection of designs from the study [by permission from CO97a].

The images were produced by decoding a binary string, U, of length 940 bits, consisting of a series of 20 binary substrings, u_i, representing each object within the image. Each object itself is defined by eight parameters, p_{ij}, which describe the type of object, its position and its colour:

$$U = u_1 \oplus u_2 \oplus u_3 \oplus \ldots\ldots\ldots \oplus u_{20}$$

and

$$u_i = p_{i1} \oplus p_{i2} \oplus p_{i3} \oplus p_{i4} \oplus p_{i5} \oplus p_{i6} \oplus p_{i7} \oplus p_{i8} \ .$$

(The operator $\oplus$ implies concatenation.) The precise meaning of the parameters depend on the object in question, as described in Tables 6.7.1 and 6.7.2. For example, the substring

$u_1 = 01010100100110001111100001011011000101101000011$

$= p_{1,1} \oplus p_{1,2} \oplus p_{1,3} \oplus p_{1,4} \oplus p_{1,5} \oplus p_{1,6} \oplus p_{1,7} \oplus p_{1,8}$

$= 01 \oplus 01010010 \oplus 01100011 \oplus 11100001 \oplus 01101100 \oplus 01011010$
$\oplus 0001 \oplus 1.$

Therefore (using Tables 6.7.1 and 6.7.2):

$p_{1,1} = 01 \Rightarrow object$ = rectangle;
$p_{1,2} = 01010010 \Rightarrow x_1 = 82$;
$p_{1,3} = 01100011 \Rightarrow y_1 = 99$;
$p_{1,4} = 11100001 \Rightarrow x_2 = 225$;
$p_{1,5} = 01101100 \Rightarrow y_2 = 108$;
$p_{1,6} = 01011010 \Rightarrow fill$ = not opaque;
$p_{1,7} = 0001 \Rightarrow colour$ = colour no. 1 (dark blue); and
$p_{1,8} = 1 \Rightarrow visible$ = true.

Thus the computer draws a visible rectangle with the above attributes as the first object (u_1). Nineteen further objects ($u_2.....u_{20}$) are then overlaid to complete the image.

The GA used was Goldberg's SGA [GO89] coded in C. Although the binary search space consists of 2^{940} ($\approx 2 \times 10^{72}$) possibilities, not all of these are distinguishable, either because of large opaque objects covering other objects coded to the left of them on the string, or because not all combinations decode differently (Table 6.7.2).

Object	p_1	p_2	p_3	p_4	p_5	p_6	p_7	p_8
line	object type	x_1	y_1	x_2	y_2	width	colour	visible
rectangle	object type	x_1	y_1	x_2	y_2	fill	colour	visible
ellipse	object type	x_1	y_1	r_2	r_2	fill	colour	visible
triangle	object type	x_1	y_1	x_2	height	fill	colour	visible

Table 6.7.1. Parameter definitions. For example, if the object is a triangle, p_5 gives the height. If the object is an ellipse, p_5 indicates the length of the minor axis [CO97a].

Term	Meaning	Integer Range	Coding	Bits
visible	whether the object is to be drawn	0 or 1	draw if 1	1
colour	colour of the object	0 to 15	-	4
width	thickness of a straight line	0-255	INT(width/20)	8
fill	whether a rectangle or ellipse is opaque	0-255	opaque if > 128	8
height	height of triangle	0-255	screen co-ordinates	8
x_1, y_1	position of the bottom left corner or centre of ellipse	0-255	screen co-ordinates	8
x_2, y_2	position of the top right corner	0-255	screen co-ordinates	8
r_1, r_2	length of major and minor axes of ellipse	0-255	screen co-ordinates	8
type	whether line, rectangle, ellipse or triangle	0-3	0=line, 1=rectangle, 2=ellipse, 3=triangle	2

Table 6.7.2. Parameter encodings used in Table 6.7.1. Triangles always have a base parallel to the bottom of the screen. Screen co-ordinates run from 0 to 255 in both x- and y-directions. If $x_2 < x_1$, or $y_2 < y_1$, then the terms left and right, or top and bottom, reverse [CO97a].

Fitness-proportional selection was used together with single point crossover and mutation to create new novel images. The GA is, in essence, providing a mechanism for the melding of information within the images, biased by the selection procedure toward those that receive higher scores. However, being a stochastic algorithm, there is no guarantee that this information will be preserved in the next generation. Mutation, for example, could force radical changes to an image.

SUBJECT TESTING

The design strategy was tested on 51 individuals (all undergraduates from various University departments). The subjects were asked to score the designs with a number between 0 and 9, the higher numbers being awarded to the most-liked designs. The fitness function (or selection pressure) applied within the GA was simply the score divided by 10. A crossover probability of 0.6 was used with a mutation probability of 0.06. These parameters were chosen to speed the search and were estimated by trial and error using additional small groups of other subjects. Each subject viewed 20 generations of 36 images, or 720 images in total. In total the test group viewed 36,720 images.

Figure 6.7.2 shows the score of the most-liked design (different for each subject) as a function of generation, averaged over all subjects. The subjects are seen to score later generations of designs higher than earlier ones. This indicates the GA is moving towards higher scoring areas of the search space at an observable rate. The linear overall form of the curve also indicates that the strategy is managing to avoid becoming lost in local minima—possibly not surprising, as so few generations are processed.

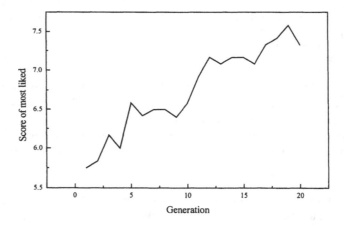

Figure 6.7.2. Score of the best design as a function of generation averaged over all subjects [CO97a].

As the scoring scale used was not absolute, it was possible that subjects were simply ranking the later images better, regardless of content. It was also possible, in fact likely, that they were not always consistent in their scoring.

Some might have been scoring against some ideal design, other simply scoring against the other images they had recently seen on the screen. An attempt was made to partially circumvent these problems by making use of a simple rank-ordering method. At the end of the test, the subjects were shown a smaller version of their most-liked design from each generation (simultaneously and randomly ordered on a single screen) and asked to rank-order them. Figure 6.7.3 shows the correlation of generation and rank-order averaged over all the subjects. This clearly shows that the images created later are preferred.

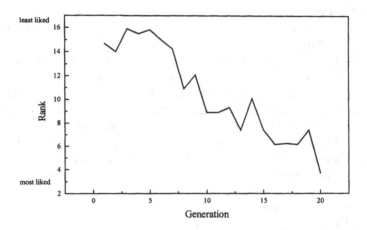

Figure 6.7.3. Final rank as a function of generation averaged over all subjects [CO97a].

One advantage of the GA is that the string representations lead themselves naturally to a (simplistic) definition of similarity and convergence. By carrying out bit-wise comparisons of the strings, across all subjects, an approximate numeric estimate of the similarity of the pictures can be made. Without a string representation, this would have to be performed on a qualitative visual basis by the researcher and is thus likely to be researcher-dependent.

The results indicate that a GA-based search can be efficient in searching for subjects' preferences of a real aesthetic product. The value placed on designs is found to rise at a useful rate with generation, and convergence is seen within the bit patterns used to represent the product. However, it is not yet clear how successful such an approach might be with other, more complex, objects.

6.8 MULTI-OBJECTIVE NETWORK REHABILITATION BY MESSY GA

The refurbishment of ageing water systems requires decisions to be made as to how best to spend a limited budget in order to maximise the level of improvement. The pre-existing network is likely to be large, complex and costly, implying that any reasonable budget will only be able to make alterations to a small number of system components. This in turn implies that most technically feasible adjustments will not be financially feasible. Thus the population of truly feasible solutions is likely to be small in comparison to the size of the search space. Such a search space, with most of the space counting as unfeasible, creates problems for any optimisation routine, including a genetic algorithm.

Godfrey Walters and Dragan Savic, working together with D. Halhal and D. Ouazar, have formulated a *structured messy genetic algorithm* (SMGA) which is particularly effective in avoiding unfeasible parts of such a space. The task examined is complicated by being a multi-objective optimisation problem requiring two factors to be simultaneously, but independently, optimised.

In the work described here, which is taken from [HA97], an SMGA is applied to both an illustrative network and a real network of 167 pipes. The work shows an example of a GA and a problem with the following features:

- a sparse problem space;
- use of a *messy* algorithm [Go89a];
- a multi-objective problem; and
- fitness sharing.

INTRODUCTION

Ageing water networks can suffer problems of water loss and reduced carrying capacity. Alongside this can run increasing consumer demand. This combination may lead to consumer discontent. However, due to budgetary and other constraints, replacement of the complete network will be unfeasible, unnecessary and possibly undesirable for other reasons. It is likely that funds will only be available to replace, rehabilitate, duplicate or repair a small number of the system components—pumps, pipes, tanks, etc.—at any one time. Hence a decision problem arises of trying to maximise the benefits of changes whilst remaining within a budgetary constraint. The optimisation problem itself is one of choosing a small number of possible designs from a much larger set of technically feasible ones.

Given a commercial operation, competition for funds means that designs which do not necessarily maximise benefits, but manage substantial improvements, may need to be considered. This implies a multiobjective problem, where a single fitness function cannot be formed. In this instance, the cost of the design must be minimised, while simultaneously the benefit of the design must be maximised. If an exact financial value could be placed upon all benefits then the problem could be collapsed to a single fitness function. However, as discussed in Chapter 4, this is often not possible.

There have been several attempts at applying algorithmic methods to network rehabilitation problems [WO87,KI94a,MU94]. Of particular relevance is the work of Murphy and Simpson [MU92] who used a genetic algorithm to find the optimal solution for a particular network. These attempts have all been based on small networks with the single objective of minimising cost subject to performance constraints, such as minimum pressure for consumers. The problem to be studied here is one of maximising benefit subject to limits on funding.

THE PROBLEM

Given a particular, pre-existing, network the desire is to invest some or all of a limited budget in order to enhance the performance of the network. In the model considered, performance may be enhanced by:

1. increasing the hydraulic capacity of the network by cleaning, relining, duplicating or replacing existing pipes;
2. increasing the physical integrity of the network by replacing structurally weak pipes;
3. increasing system flexibility by adding additional pipe links; and
4. improving water quality by removing or relining old pipes.

The problem can be stated as:

$$\text{maximise } Benefit(i)$$

and

$$\text{minimise } Cost(i)$$

subject to

$$Cost \leq Budget$$

where i represents a particular solution.

In order to allow the GA to be able to preferentially select one solution over another, *Benefit* must be able to assume a numerical value. This is done by forming a weighted sum based on the four factors *hydraulic capacity, integrity, flexibility* and *quality* described above. Thus the *Benefit* of solution *i* is defined as:

$$Benefit(i) = w_h B_h(i) + w_p B_p(i) + w_f B_f(i) + w_q B_q(i) \; , \qquad (6.8.1)$$

where B_h, B_p, B_f, and B_q describe the hydraulic, physical, flexibility and quality benefits respectively and w_h, w_p, w_f, and w_q their respective weights (which are user defined).

Each of the four benefits are defined as follows (for more details see [HA97]):

- The hydraulic benefit, B_h, is defined as the reduction of the level of deficiency (caused by pressure shortfalls) allowed by adoption of a particular strategy. (The nodal pressures are estimated by use of the steady-state hydraulic network solver EPANET [ROS93].)
- Improvements to the structural condition of the network pipes reduce future repair costs, the sum of which gives B_p.
- The laying of duplicate pipes in parallel with pre-existing ones increases the flexibility of the system, with benefit B_f, proportional to the number of pipes replaced.
- A pipe with a low Hazen-Williams (HW) factor is probably suffering from corrosion, tuberculation or scaling—all of which can help the development of micro-organisms or lead to discoloured water. The replacement of such pipes can therefore lead to a water quality benefit, B_q, proportional to the length of replaced pipe.

THE STRUCTURED MESSY GENETIC ALGORITHM (SMGA) APPROACH

Messy GAs are an attempt to allow for the progressive growth in the complexity of a solution by allowing the chromosome to increase in length with time. This is achieved by repetitive application of a GA and the concatenation of strings representing partial solutions. In this respect the method imitates long-term evolution of single-cell organisms to complex life-forms such as ourselves. It is an extremely interesting approach and allows more traditional GAs to be seen as simple adapting algorithms that in essence fine tune the fitness of pre-existing *species*. With a messy GA these species have first to be built before they are adapted to their environment. The SMGA introduced by Halhal, Walters, Ouazar and Savic [HA97] proceeds as follows:

<u>Step 1</u> Enumerate <u>ALL</u> single-variable decisions on the network—for example, "replace pipe 89". The point to note is that only a single decision is being made. This implies that a great number (if not all) of such decisions are feasible, in that their cost will be less than the budget. This is one of the strengths of the approach for the type of problem considered here: right from the start feasible solutions are processed. In a sparse search space, with few feasible solutions, this would not be so for a more typical GA.

Because population members need to describe not only the value of a variable but also which variable is being described, a coding scheme is required such that an *element* both identifies the design variable and its value. This first stage is then the enumeration of all possible single elements.

<u>Step 2</u> Filter the population to remove less well performing individuals. The remaining individuals are retained to provide subsequent populations with high performing elemental building blocks.

<u>Step 3</u> Increase the complexity of population members by adding a single elemental building block to each member (concatenation).

<u>Step 4</u> Use this population as the initial population of a GA and run the algorithm until a termination criterion is met.

<u>Step 5</u> Repeat from step 3, unless either the strings have reached a predetermined length, or no improvement to the solution is seen for a set number of successive concatenation steps.

As a string within the SMGA generally only contains a small number of possible decision variables, its length is typically much less than that required to contain all decision variables. This is despite the need to hold information about the variable values as well as a tag to identify which variables are being described. Not only does this improve the use of computer memory, but more importantly, it partially circumvents the slow progress made by standard GAs containing very long strings.

If typically only p network arcs (pipes) can be considered out of a total of q arcs because of budgetary constraints, then if each fragment can take n alternative solutions, the search space contains:

$$n^p \frac{q!}{p!(q-p)!}$$

(6.8.2)

solutions [HA97]; whereas it contains

$$n^p \qquad\qquad (6.8.3)$$

solutions for a standard GA. If $q = 60$, $p = 6$, and $n = 4$, then (6.8.2) implies 2.7×10^{12} possibilities, a seemingly large space, until it is realised that (6.8.3) implies 1.3×10^{36} possibilities.

USING THE SMGA FOR MULTIOBJECTIVE OPTIMISATION

Given more than one objective, it is impossible to form a simple fitness function of the type used in earlier examples. One way around this would be to use the second objective as a constraint—for example, by hunting for solutions which provide a set minimum benefit. Another possibility would be to factor benefits into the cost objective, by assigning benefit a financial cost. Neither of these approaches is entirely satisfactory.

True multi-objective optimisation treats each objective separately. This implies that each solution is not a single point in the fitness landscape, but a vector with one dimension for each objective. The Pareto optimal set of such vectors are those solutions which are undominated by any other solutions (see Chapter 4). For such solutions it is impossible to improve one objective without simultaneously making one or more of the other objectives worse.

It is relatively easy to use the idea of a Pareto optimal set to drive selection within a GA. Typically, this is achieved by finding all current members of the Pareto optimal set within the population, assigning them the highest rank and removing them from the population—thus creating a new Pareto optimal set (or *front*) which is assigned the next highest rank. This process is repeated until all members of the population are ranked. Rank-order selection is then used (together with crossover and mutation) to build the next generation. Although simple in outline, caution is required to avoid convergence to a single solution.

As stressed by Halhal et al. [HA97], the SMGA is naturally suited to multiobjective problems because it ensures a good spread of different solutions across the range of feasible costs. A standard GA is unlikely to do this because, within a sparse solution space, many (if not most) of the members of early generations will represent unfeasible solutions.

For the water rehabilitation problem, the fitness f_i of each individual i was assigned through:

$$f_i = \frac{1}{\text{rank(i)}} \; .$$

FITNESS SHARING

In order to preserve a reasonable spread of solutions along each Pareto optimal front Halhal et al. [HA97] included fitness sharing [GO87,GO89,DE89] (see Chapter 4) in the algorithm. This reduces the level of competition between similarly ranked, but distant, population members. Niche formation and speciation are achieved by dividing the budget into a series of intervals or *classes*. Each individual is assigned to the particular interval that includes the cost of the particular solution.

A class is deemed *full* if it contains N/n individuals, where n is the number of classes. The shared fitness f_i^{share} of individual i is given by

$$f_i^{share}(i) = \frac{f_i}{N_c(j)} \; ,$$

where $N_c(j)$ is the number of individuals in class j. The value of n is, itself, a function of generation.

SMGA PARAMETERS

An integer coding is used. As discussed above, both the decision and the decision variable must be held in the string. This is achieved by using a substring to represent all the decision variables (the arc numbers) and a second substring to hold the decisions (what is done to each arc). For example, a possible string might be:

arc-substring

$$\overbrace{2\ 1\ 5\ 3\ 6}\ \ 1\ 8\ 2\ 4\ 2$$

$$\underbrace{}$$

decision-substring

During the concatenation phase (step 3, above) a single digit (decision) is added to each substring in the population. Highly performing digits from the initial population are used unless this causes a duplicate, in which case a random digit is used. The strings (now of identical length) then undergo processing via a conventional GA.

N is set to 40, with n ($g = 0$) set to 5. Two-point crossover is applied separately to each substring, with child arc-substring being checked to ensure no arcs are duplicated. Any duplications are again replaced by arcs chosen at random.

APPLICATIONS

The SMGA was applied to the network depicted in Figure 6.8.1. The network consists of 15 pipes, nine nodes and seven loops. There are eight possible decisions for each pipe: "leave as is"; "clean and line"; "duplicate" (with one of four set diameters); "renew with same diameter"; or "renew with next largest diameter". The available fund is assumed to be 2,000,000.

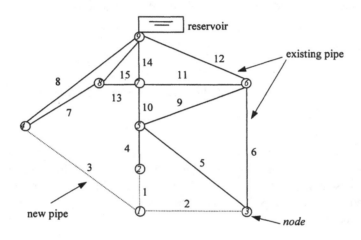

Figure 6.8.1. The test network (node numbers shown in italics, pipes in roman) [HA97].

The algorithm was applied with and without fitness sharing for a total of 5,000 objective function evaluations, three times. Fitness sharing was found to produce a slightly more even spread of solutions across the range of possible costs. The performance of SMGA (with sharing) was compared with an integer-coded standard GA (with sharing) where the string length equalled the number of arcs (pipes) in the network and the integer value in each string position defined the action to be taken for the corresponding arc. The best solutions discovered in each generation with costs less than 1.1 times the budget are shown in Figure 6.8.2. SMGA produced a slightly better set of nondominated solutions. Table 6.8.1 lists a selection of the nondominated solutions (see §4.4) identified by SMGA showing which arcs (pipes) should be replaced and which left.

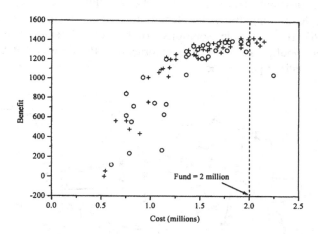

Figure 6.8.2. Best solutions from each generation; + SMGA, o standard-GA [HA97].

Cost	Benefit	Min. Pressure	Arc 1	Arc 2	Arc 3	Arc 4	Arc 6	Arc 13	Arc 14	Arc 15
1940000	1433.72	7.47	80	80	150	L	L	300ᴾ	L	400ᴾ
1911000	1428.96	10.28	150	80	150	L	L	L	300ᴾ	200ᴾ
1780000	1417.67	5.85	80	80	150	L	L	L	400	L
1755000	1414.81	3.82	80	80	150	L	L	L	300ᴾ	200ᴾ
1585000	1388.83	1.00	80	80	80	150ᴾ	L	L	300ᴾ	L
1435000	1369.52	-2.47	80	80	150	L	L	L	300ᴾ	L
1380000	1311.64	-7.06	80	80	150	L	L	150ᴾ	L	300ᴾ
1367500	1305.26	-15.00	80	80	100	L	L	L	300ᴾ	L
1255000	1260.33	-27.38	80	80	80	L	L	L	300ᴾ	L
1160000	1125.96	-13.54	80	80	150	L	L	L	L	300ᴾ
930000	1030.88	-30.02	80	80	150	L	L	L	L	200ᴾ
750000	848.15	-60.56	80	80	80	L	L	L	L	200ᴾ

Table 6.8.1. Nondominated solutions for the test network, including new pipe diameters (ᴾ add parallel pipe of diameter shown, L leave) [HA97].

As a second, and more realistic, example Halhal *et. al.* applied the SMGA to the network shown in Figure 6.8.3 and really demonstrated the advantages of the method. This is the water distribution of a real town of 50,000 inhabitants in Morocco. It contains 167 pipes and 115 nodes. The total number of possible solutions (assuming an unlimited string length is 6.55×10^{150}. Restricting the string length to a more realistic 40 digits still implies 8.67×10^{74} possible solutions. Halhal *et. al.* suggest a complete enumeration would take 2.75×10^{62} centuries of CPU time at 1,000 objective function evaluations per second.

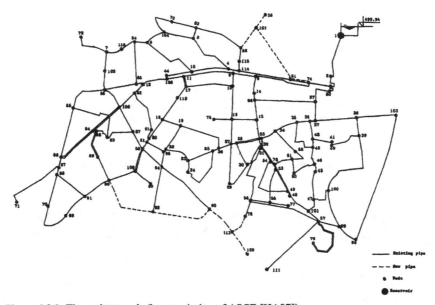

Figure 6.8.3. The real network (by permission of ASCE [HA97]).

The benefits of the SMGA over a standard GA of fixed string length were gauged by carrying out three independent runs of each. 25,000 objective function evaluations were allowed in each case. The standard GA (Goldberg's SGA [GO89]) failed to converge to a design with costs less than the budget. Even the use of a high penalty function, in order to reduce the chance of high-cost solutions being reproduced, was only a partial improvement. It still took 7,000 objective function evaluations for the standard GA to converge to a feasible solution. Furthermore, subpopulations in only four out of eight cost classes were formed. In contrast, the SMGA formed subpopulations in all cost classes from the beginning (Figure 6.8.4).

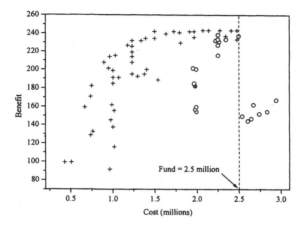

Figure 6.8.4. Best solutions of each generation for the SMGA (+) and a standard GA (○) (data from [HA97]).

Appendix A. Electronic- and Paper-Based Resources

The are many electronic- and paper-based resources available about GAs. The internet is also an ideal way of finding those that have tried to solve similar problems, or used similar methods.

Electronic-Based Resources

The following two world-wide web sites contain a multitude of information, software and pointers to other GA sites:

> http://alife.santafe.edu

> http://www.aic.nrl.navy.mil/galist

A FORTRAN GA code is available from David Carroll's web site:

> http://www.staff.uiuc.edu/~carroll/ga.html

Paper-Based Resources

Apart from the text books mentioned at the beginning of Chapter 6, there are several other general references which contain a high density of information: in particular, the series *Foundations of Genetic Algorithms* and the *Proceedings of the International Conference on Genetic Algorithms*:

Grefenstette, J.J, (Ed.), *Proceedings of an International Conference on Genetic Algorithms and their Applications*, Morgan Kaufmann, 1985.

Grefenstette, J.J, (Ed.), *Genetic Algorithms and their Applications: Proceedings of the Second International Conference on Genetic Algorithms*, Erlbaum, 1987.

Schaffer, J.D., (Ed.), *Proceedings of the Third International Conference on Genetic Algorithms*, Morgan Kaufmann, 1989.

Belew, R.K. and Booker, L.B., (Eds.), *Proceedings of the Fourth International Conference on Genetic Algorithms*, Morgan Kaufmann, 1991.

Forrest, S., (Ed.), *Proceedings of the Fifth International Conference on Genetic Algorithms*, Morgan Kaufmann, 1993.

Eshelman, L.J., (Ed,), *Proceedings of the Sixth International Conference on Genetic Algorithms*, Morgan Kaufmann, 1995.

Bäck, T., (Ed,), *Proceedings of the Seventh International Conference on Genetic Algorithms*, Morgan Kaufmann, 1997.

Rawlins, G., (Ed.), *Foundations of Genetic Algorithms*, Morgan Kaufmann, 1991.

Whitley, D., (Ed.), *Foundations of Genetic Algorithms 2*, Morgan Kaufmann, 1993.

Whitley, D. and Vose, M. (Eds.), *Foundations of Genetic Algorithms 3*, Morgan Kaufmann, 1995.

Belew, R.K. and Vose, M. (Eds.), *Foundations of Genetic Algorithms 4*, Morgan Kaufmann, 1997.

APPENDIX B. A COMPLETE LISTING OF LGADOS.BAS

```
'LGADOS - A DOS based version of the LGA Genetic Algorithm.

'For Distribution with the book "An Introduction to Genetic Algorithms for
'Scientists and Engineers", World Scientific 1998.

'David A. Coley
'Complex Systems Group
'Physics Building
'University of Exeter
'Exeter
'EX4 4QL
'UK
'email D.A.Coley@exeter.ac.uk

'
```

```
'----- DECLARE ALL THE SUBROUTINES (PROCEDURES) USED BY THE PROGRAM

DECLARE SUB OpenFiles ()
DECLARE SUB Scaling (ScalingConstant!, FittestIndividual!, SumFitness!, MeanFitness!)
DECLARE SUB Elite (SumFitness!, FittestIndividual!)
DECLARE SUB Selection (mate!, SumFitness!, MeanFitness!)
DECLARE SUB CrossOver (Mate1!, Mate2!, NewIndividual!)
DECLARE SUB FindFitness ()
DECLARE SUB PrintGeneration (Generation, MeanFitness!, FittestIndividual!)
```

```
DECLARE SUB DefineRange ()
DECLARE SUB FindIntegers ()
DECLARE SUB FindUnknowns ()
DECLARE SUB InitialPopulation ()
DECLARE SUB NoCrossover (Mate1!, Mate2!, NewIndividual!)
DECLARE SUB Mutate ()
DECLARE SUB Replace ()
DECLARE SUB Statistics (MeanFitness!, SumFitness!, FittestIndividual!, Generation)
```

'——— *SET ALL THE IMPORTANT FIXED PARAMETERS.* ———

These should be set by the user.

```
CONST PopulationSize = 20   'Must be even.
CONST NumberOfUnknowns = 2
CONST SubstringLength = 12   'All sub-strings have the same length.
CONST TotalStringLength = NumberOfUnknowns * SubstringLength
CONST MaxGeneration = 20  'G.
CONST CrossOverProbability = .6    'Pc. >=0 and <=1.
CONST MutationProbability = 1 / TotalStringLength    'Pm, >=0 and <1.
CONST Elitism = "on"  '"on" or "off".
CONST ScalingConstant = 1.2  'A value of 0 implies no scaling.
```

'———*DECLARE ALL SHARED (IE. GLOBAL) VARIBLES*———.

The arrays that hold the individuals within the current population.

```
DIM SHARED Unknowns (PopulationSize, NumberOfUnknowns) AS SINGLE
DIM SHARED Integers (PopulationSize, NumberOfUnknowns) AS LONG
DIM SHARED Strings (PopulationSize, TotalStringLength) AS INTEGER
DIM SHARED Fitness (PopulationSize) AS SINGLE
```

The new population.

```
DIM SHARED NewStrings(PopulationSize, TotalStringLength) AS INTEGER

'The array that defines the range of the unknowns.
DIM SHARED Range(2, NumberOfUnknowns) AS SINGLE

'The best individual in the past generation. Used if elitism is on.
DIM SHARED EliteString(TotalStringLength) AS INTEGER
DIM SHARED EliteIntegers(NumberOfUnknowns) AS LONG
DIM SHARED EliteFitness AS SINGLE
DIM SHARED EliteUnknowns(NumberOfUnknowns) AS SINGLE

CLS      'Clear the screen.

CALL DefineRange     'Define the range of each unknown. These should also be set by the user.

'Set the random number generator so it produces a different set of numbers
'each time LGADOS is run.
RANDOMIZE TIMER

CALL OpenFiles   'Open files used to store results.

'—— START OF THE GENETIC ALGORITHM ——

'—— CREATE AN INITIAL POPULATION, GENERATION 1 ——

Generation = 1

CALL InitialPopulation   'Build a population of strings at random.

CALL FindFitness   'Find the fitness of each member of the population.
```

```
CALL Statistics (MeanFitness, SumFitness, FittestIndividual, Generation)   'Find the mean fitness and the fittest
                                                                           'individual.

CALL PrintGeneration (Generation, MeanFitness, FittestIndividual)   'Print generation to file.

CALL Scaling (ScalingConstant, FittestIndividual, SumFitness, MeanFitness)   'If linear fitness scaling is "on"
                                                                            'then scale population prior to selection.

'------ LOOP OVER ALL THE GENERATIONS ------

FOR Generation = 2 TO MaxGeneration

  FOR NewIndividual = 1 TO PopulationSize STEP 2   'Loop over the population choosing pairs of mates.

    CALL Selection (Mate1, SumFitness, MeanFitness)   'Choose first mate.
    CALL Selection (Mate2, SumFitness, MeanFitness)   'Choose second mate.

    'Pass individuals to the temporary population either with or without performing crossover.
    IF RND <= CrossOverProbability THEN   'Perform crossover.
      CALL CrossOver (Mate1, Mate2, NewIndividual)
    ELSE 'Don't perform crossover.
      CALL NoCrossover (Mate1, Mate2, NewIndividual)   'Don't perform crossover.
    END IF

  NEXT NewIndividual

  CALL Mutate   'Mutate the temporary population.

  CALL Replace   'Replace the old population completely by the new one.
```

```
CALL FindUnknowns    'De-code the new population to integers then real numbers.

CALL FindFitness    'Find the fitness of each member of the population.

CALL Statistics(MeanFitness, SumFitness, FittestIndividual, Generation)    'Find the mean fitness and the 'fittest
                                                                             individual .

CALL PrintGeneration(Generation, MeanFitness, FittestIndividual)    'Print generation to file.

CALL Scaling(ScalingConstant, FittestIndividual, SumFitness, MeanFitness)    'If linear fitness scaling is "on"
                                                                             'then scale population prior to selection.

NEXT Generation    'Process the next generation.

CLOSE    'Close all files

SUB CrossOver (Mate1, Mate2, NewIndividual)
'Perform single point crossover.

CrossSite = INT((TotalStringLength - 1) * RND + 1)    'Pick the cross-site at random.

FOR Bit = 1 TO CrossSite    'Swap bits to the left of the cross-site.
    NewStrings(NewIndividual, Bit) = Strings(Mate1, Bit)
    NewStrings(NewIndividual + 1, Bit) = Strings(Mate2, Bit)
NEXT Bit

FOR Bit = CrossSite + 1 TO TotalStringLength    'Swap bits to the right of the cross-site.
    NewStrings(NewIndividual, Bit) = Strings(Mate2, Bit)
    NewStrings(NewIndividual + 1, Bit) = Strings(Mate1, Bit)
NEXT Bit
```

```
END SUB

SUB DefineRange
'Defines the upper and lower bounds of each unknown.
'Add other ranges using the same format if more than two unknowns in the problem.

Unknown = 1   'the first unknown.
Range(1, Unknown) =  0   'The lower bound.
Range(2, Unknown) =  1   'The upper bound.

Unknown = 2   'the second unknown.
Range(1, Unknown) = -3.14159
Range(2, Unknown) =  3.14159

'Add other ranges if more than two unknowns in your problem.

END SUB

SUB Elite (SumFitness, FittestIndividual)
'Applies elitism by replacing a randomly chosen individual by the elite member
'from the previous population if the new max fitness is less then the previous value.

IF Fitness(FittestIndividual) < EliteFitness THEN

  Individual = INT(PopulationSize * RND + 1)   'Chosen individual to be replaced.

  FOR Bit = 1 TO TotalStringLength
    Strings(Individual, Bit) = EliteString(Bit)
  NEXT Bit
```

```
      Fitness(Individual) = EliteFitness

      FOR Unknown = 1 TO NumberOfUnknowns
         Integers(Individual, Unknown) = EliteIntegers(Unknown)
         Unknowns(Individual, Unknown) = EliteUnknowns(Unknown)
      NEXT Unknown

      FittestIndividual = Individual

   END IF

   FOR Bit = 1 TO TotalStringLength
         EliteString(Bit) = Strings(FittestIndividual, Bit)
   NEXT Bit

   EliteFitness = Fitness(FittestIndividual)

   FOR Unknown = 1 TO NumberOfUnknowns
      EliteIntegers(Unknown) = Integers(FittestIndividual, Unknown)
      EliteUnknowns(Unknown) = Unknowns(FittestIndividual, Unknown)
   NEXT Unknown

END SUB

SUB FindFitness
```

The problem at hand is used to assign a positive (or zero) fitness to each individual in turn.

The problem is $f = x^2 + sin(y)$.

```
FOR Individual = 1 TO PopulationSize
   Fitness(Individual) = Unknowns(Individual, 1) ^ 2 + SIN(Unknowns(Individual, 2))
   If Fitness(Individual) < 0 then Fitness(Individual) = 0
```

```
NEXT Individual

END SUB

SUB FindIntegers
'Decode the strings to sets of decimal integers.

DIM bit AS INTEGER

FOR Individual = 1 TO populationsize
    bit = TotalStringLength + 1
    FOR Unknown = NumberOfUnknowns TO 1 STEP -1
        Integers(Individual, Unknown) = 0
        FOR StringBit = 1 TO SubstringLength
            bit = bit - 1
            IF Strings(Individual, bit) = 1 THEN
                Integers(Individual, Unknown) = Integers(Individual, Unknown) + 2 ^ (StringBit - 1)
            END IF
        NEXT StringBit
    NEXT Unknown
NEXT Individual

END SUB

SUB FindUnknowns
'Decode the strings to real numbers.

CALL FindIntegers   'First decode the strings to sets of decimal integers.

'Now convert these integers to reals.
FOR Individual = 1 TO PopulationSize
```

```
      FOR Unknown = 1 TO NumberOfUnknowns
        Unknowns(Individual, Unknown) = Range(1, Unknown) + Integers(Individual, Unknown) * (Range(2,
                                        Unknown) - Range(1, Unknown)) / (2 ^ SubstringLength - 1)

      NEXT Unknown
    NEXT Individual

END SUB

SUB InitialPopulation
'Create the initial random population.

    FOR Individual = 1 TO PopulationSize

      FOR Bit = 1 TO TotalStringLength
        IF RND > .5 THEN
          Strings(Individual, Bit) = 1
        ELSE
          Strings(Individual, Bit) = 0
        END IF
      NEXT Bit

    NEXT Individual

    CALL FindUnknowns   'Decode strings to real numbers.

END SUB

SUB Mutate
'Visit each bit of each string very occasionally flipping a "1" to a "0" or vice versa.

    FOR Individual = 1 TO PopulationSize
```

```
FOR Bit = 1 TO TotalStringLength

    'Throw a random number and see if it is less than or equal to the mutation probability.
    IF RND <= MutationProbability THEN

        'Mutate.
        IF NewStrings(Individual, Bit) = 1 THEN
            NewStrings(Individual, Bit) = 0

        ELSE
            NewStrings(Individual, Bit) = 1

        END IF

    END IF

NEXT Bit

NEXT Individual

END SUB

SUB NoCrossover (Mate1, Mate2, NewIndividual)
'Pass the selected strings to the temporary population without applying crossover

FOR Bit = 1 TO TotalStringLength
    NewStrings(NewIndividual, Bit) = Strings(Mate1, Bit)
    NewStrings(NewIndividual + 1, Bit) = Strings(Mate2, Bit)
NEXT Bit

END SUB

SUB OpenFiles
```

```
'Open result files. See Chapter 2 for a description of their contents.

OPEN "LGADOS.RES" FOR APPEND AS #1
OPEN "LGADOS.ALL" FOR OUTPUT AS #2

END SUB

SUB PrintGeneration (Generation, MeanFitness, FittestIndividual)
'Print results to the screen and the files.

PRINT Generation; Fitness(FittestIndividual); MeanFitness;        'Screen.
PRINT #1, Generation; ","; Fitness(FittestIndividual); ","; MeanFitness;    'File LGADOS.RES.

FOR Unknown = 1 TO NumberOfUnknowns
    PRINT Unknowns(FittestIndividual, Unknown);                   'Screen.
    PRINT #1, ","; Unknowns(FittestIndividual, Unknown);    'File LGADOS.RES
NEXT Unknown

PRINT                       'Carriage return.
PRINT #1,                   'Carriage return.

FOR Individual = 1 TO populationsize

    PRINT #2, Generation; ","; Fitness(Individual); ","; 'File LGADOS.ALL

    FOR Unknown = 1 TO NumberOfUnknowns
        PRINT #2, Unknowns(Individual, Unknown); ","; 'File LGADOS.ALL
    NEXT Unknown

    FOR Bit = 1 TO TotalStringLength
```

```
    PRINT #2, RIGHT$(STR$(strings(Individual, Bit)), 1) ;","; 'File LGADOS.ALL
  NEXT Bit

  PRINT #2, 'Carriage return

NEXT Individual

END SUB

SUB Replace
'Replace the old population with the new one.

FOR Individual = 1 TO PopulationSize
  FOR Bit = 1 TO TotalStringLength
    strings(Individual, Bit) = NewStrings(Individual, Bit)
  NEXT Bit
NEXT Individual

ERASE NewStrings    'Clear the old array of strings.

END SUB

SUB Scaling (ScalingConstant, FittestIndividual, SumFitness, MeanFitness)
```

'Apply Linear Fitness Scaling,
'   scaledfitness = a * fitness + b.
'Subject to,
'   meanscaledfitness = meanfitness
'and
'   bestscaledfitness = c * meanfitness,
'where c, the scaling constant, is set by the user.

'If the scaling constant is set to zero, or all individuals have the same fitness, scaling is not applied.

```
IF ScalingConstant <> 0 AND Fitness(FittestIndividual) - MeanFitness > 0 THEN
```

'Find a and b.

```
    a = (ScalingConstant - 1) * MeanFitness / (Fitness(FittestIndividual) - MeanFitness)

    b = (1 - a) * MeanFitness
```

'Adjust the fitness of all members of the population.

```
    SumFitness = 0
    FOR Individual = 1 TO PopulationSize
        Fitness(Individual) = a * Fitness(Individual) + b
        IF Fitness(Individual) < 0 THEN Fitness(Individual) = 0  'Avoid negative values near the end of a run.
        SumFitness = SumFitness + Fitness(Individual)  'Adjust the sum of all the fitnesses.
    NEXT Individual
```

'Adjust the mean of all the fitnesses.

```
    MeanFitness = SumFitness / PopulationSize
END IF

END SUB

SUB Selection (mate, SumFitness, MeanFitness)
```

'Select a single individual by fitness proportional selection.

```
Sum = 0
Individual = 0

RouletteWheel = RND * SumFitness
```

```
DO
    Individual = Individual + 1
    Sum = Sum + Fitness(Individual)
LOOP UNTIL Sum >= RouletteWheel OR Individual = PopulationSize

mate = Individual

END SUB

SUB Statistics (MeanFitness, SumFitness, FittestIndividual, Generation)
'Calculate the sum of fitness across the population and find the best individual,
'then apply elitism if required.

FittestIndividual = 0
MaxFitness = 0

FOR Individual = 1 TO PopulationSize
    IF Fitness(Individual) > MaxFitness THEN
        MaxFitness = Fitness(Individual)
        FittestIndividual = Individual
    END IF
NEXT Individual

IF Elitism = "on" THEN 'Apply elitism.
    CALL Elite(SumFitness, FittestIndividual)
END IF

SumFitness = 0 'Sum the fitness.
FOR Individual = 1 TO PopulationSize
    SumFitness = SumFitness + Fitness(Individual)
NEXT Individual
```

```
'Find the average fitness of the population.
MeanFitness = SumFitness / PopulationSize

END SUB
```

REFERENCES

AL77   Alperovits, E and Shamir, U, Design of optimal water distribution systems, *Water Resour. Res.* **13**(6), p885-900, 1977.

AL95   Altenberg, L., The schema theorem and Price's theorem, Whitley, L.D. and Vose, M., (Eds.), *Foundations of Genetic Algorithms 3*, Morgan Kaufmann, 1993.

AN89   Antonisse, J., A new interpretation of schema notation that overturns the binary encoding constraint, in Schaffer, J.D., (Ed.), *Proceedings of the 3rd International Conference on Genetic Algorithms*, Morgan Kaufmann, p86-91, 1989.

AR89   Arora, J.S., *Introduction to Optimum Designs*, McGraw-Hill, 1989.

BA85   Baker, J.E., Adaptive selection methods for genetic algorithms, *Proceedings of an International conference on Genetic Algorithms and their Applications*, p101-111, 1985.

BA87   Baker, J. E., Reducing bias and inefficiency in the selection algorithm, in *Genetic Algorithms and their Applications: Proceedings of the Second International Conference on Genetic Algorithms*, 1987.

BA91   Bäck, T., Hoffmeister, F. and Schwefell, H., A survey of evolution strategies, in Belew, R.K. and Booker, L.B., (Eds), *Proceedings of the 4th International Conference on Genetic Algorithms*, Morgan Kaufmann, p2-9, 1991.

BA93   Bäck, T., Optimal mutation rates in genetic search, in *Genetic Algorithms: Proceedings of the 5th International Conference*, Forrest, S., (Ed.), p2-8, Morgan Kaufmann, 1993.

BA96   Bäck, T., *Evolutionary Algorithms in Theory and Practice*, Oxford University Press, New York, 1996.

BE93   Bertoni, A. and Dorigo, M., Implicit parallelism in genetic algorithms, *Artificial Intelligence*, **61**(2), p307-314, 1993.

BE93a   Beasley, D., Bull, D.R. and Martin, R.R., A sequential niche technique for

206

multimodal function optimization, *Evolutionary Computation*, 1(2), p101-125, 1993.

BI86   Binder, K. and Young, A.R., *Rev. Mod. Phys.* **58**, p801, 1986.

BL95   Blickle, T. and Thiele, L., A mathematical analysis of tournament selection, in Eshelman, L.J., *Proceedings of the 6th International Conference on Genetic Algorithms*, p506-511, 1995.

BO87   Booker, L., Improving search in genetic algorithms, in [DA87], p61-73, 1987.

BR89   Bramlette, M.F. and Cusic, R., A comparative evaluation of search methods applied to the parametric design of aircraft, Schaffer, J.D., (Ed.), *Proceedings of the 3rd International Conference on Genetic Algorithms*, Morgan Kaufmann, 1989.

BR91   Bramlette, M.F., Initialization, mutation and selection methods in genetic algorithms for function optimisation, in Belew, R.K. and Booker, L.B., (Eds), *Proceedings of the 4th International Conference on Genetic Algorithms*, Morgan Kaufmann, 1991.

BU84   Bunday, B.D., *Basic Optimisation Methods*, Edward Arnold, London, 1984.

CA89   Caruana, R.A., Eshelman, L.J. and Schaffer, J.D., Representation and hidden bias II: estimating defining length bias in genetic search via shuffle crossover, in *Proceedings of the 11th international Joint Conference on Artificial Intelligence*, Morgan Kaufmann, San Mateo, p750-755, 1989.

CA91   Caldwell, C. and Johnston, V.S., Tracking a Criminal Suspect Through "Face-Space" with a Genetic Algorithm, in Belew, R.K. and Booker, L.B., (Eds), *Proceedings of the 4th International Conference on Genetic Algorithms*, Morgan Kaufmann, p416-421, 1991.

CA96a   Carroll, D.L., Chemical Laser Modeling with Genetic Algorithms, *AIAA Journal*,. **34**(2), pp. 338-346, February 1996.

CA96b   Carroll, D.L., Genetic Algorithms and Optimizing Chemical Oxygen-Iodine Lasers, *Developments in Theoretical and Applied Mechanics*, Vol. XVIII, eds. Wilson, H.B., Batra, R.C., Bert, C.W., Davis, A.M.J., Schapery, R.A., Stewart, D.S. and Swinson, F.F., School of Engineering, The University of Alabama, , pp.411-424, 1996.

CE93   Celik, T., Hausmann U.H.E. and Berg B., *Computer Simulation Studies in Condensed Matter VI*, Landau, D.P., Mon, K.K., and Schuttler, H.B., (Eds.), Springer Verlag, Heidelberg, p173, 1993.

CH90   Chalmers, D.J., The evolution of learning: an experiment in genetic connectionism, in Touretzky, D.S., Elman, J.L., Sejnowski, T.J. and Hinton, G.E., (Eds.), *Proceedings of the 1990 Connectionist Models Summer School*, Morgan Kaufmann, 1990.

CH96   Chipperfield, A. and Fleming, P., Genetic algorithms in control systems engineering, *J. of Computers and Control*, **24**(1), 1996.

CH97   Chen, Y.W., Nakao, Z., Arakaki, K., Tamura, S., Blind deconvolution based on genetic algorithms, *IEICE Transactions on Fundamentals of Electronics, Communications and Computer Sciences*, **E80A**(12), p2603-2607, 1997.

CO92   Coley, D.A. and Penman, J.M., Second order system identification in the thermal response of a working school: Paper II. *Building and Environment* **27**(3) 269-277, 1992.

CO92a   Collins, R.J. and Jefferson, D.R., The evolution of sexual selection and female choice, Varela, F. J., Bourgine, P., (Eds.), *Toward a Practice of Autonomous Systems: Proceedings of the First European Conference on Artificial Life*, MIT Press, 1992.

CO92b   Cooper, D.E. (Ed.), *A Companion to Aesthetics*, Blackwell, 1992.

CO94   Coley, D.A. and Bunting, U., The identification of complex objects from NMR Images by Genetic Algorithm, *IEE Digest 193* p91-96, 1994.

CO94a   Coley, D.A. and Crabb, J.A.. Computerised Control of artificial light for maximum use of daylight, *Lighting Res. Technol.* **26** (4) p189-194, 1994.

CO96   Coley, D.A., Genetic Algorithms, *Contemporary Physics*, **37**(2) p145-154, 1996.

CO97   Coley, D.A. and Crabb, J.A., *An artificial intelligence approach to the prediction of natural lighting levels*, Building and Environment, **32**(4), p81-85, 1997.

208

CO97a Coley, D.A. and Winters, D., *Search Efficacy in Aesthetic Product Spaces*, Complexity, **3**(2), p23-27, 1997.

COH91 Cohoon, J.P., Hegde, S.U., Martin, W.N. and Richards, D.S., Distributed genetic algorithms for the floorplan design problem, *IEEE Trans., CAD*, **10**(4), p483-492, 1991.

CR84   Crozier, R.W. and Chapman, A.J. (Eds.), *Cognitive Processes in the Perception of Art*, North-Holland, 1984.

CR87   Crabb, J.A., Murdoch, N. and Penman, J.M., Validation study of EXCALIBUR, a simplified thermal response model, *Building Services Research and Technology* **8** p13-19, 1987.

CR87a Crabb, J.A., Murdoch, N. and Penman, J.M., Building energy assessment by simplified dynamic simulation model, presented at the European Conference on Architecture, Munich, April 1987.

DA87   Davis, L. (Ed.), *Genetic algorithms and simulated annealing*, Pitman, London, 1987.

DA89   Davis, L., Adapting operator probabilities in genetic algorithms, in Schaffer, J.D., (Ed.), *Proceedings of the 3rd International Conference on Genetic Algorithms*, Morgan Kaufmann, p61-69, 1989.

DA91   Davis, L., (Ed.), *Handbook of Genetic Algorithms*, Van Nostrand Reinhold, New York, 1991.

DA91a Davidor, Y., A naturally occurring niche and species phenomenon: the model and first results, in Belew, R.K. and Booker, L.B., (Eds), *Proceedings of the 4th International Conference on Genetic Algorithms*, Morgan Kaufmann, p257-263, 1991.

DA91b Davis, L., Bit-climbing, representational bias and test suit design, in Belew, R.K. and Booker, L.B., (Eds), *Proceedings of the 4th International Conference on Genetic Algorithms*, Morgan Kaufmann, p18-23, 1991.

DE75   De Jong, K.A., *Analysis of the behaviour of a class of genetic adaptive systems*,

Doctoral dissertation, University of Michigan, Dissertation Abstracts International 36(10), 5140B, (University Microfilms No. 76-9381), 1975.

DE89   Deb, K., *Genetic Algorithms in multi-modal function optimisation*, Masters thesis, The Centre for Genetic Algorithms Report No. 89002, University of Alabama, 1989.

DE89a   Deb, K., and Goldberg, D.E., An investigation of niche and species formation in genetic function optimisation, in Schaffer, J.D., (Ed.), *Proceedings of the 3rd International Conference on Genetic Algorithms*, Morgan Kaufmann, 1989.

DE93   De Jong, K. A., Genetic algorithms are NOT function optimisers, In Whitley, L.D., (Ed.), *Foundations of Genetic Algorithms 2*, Morgan Kaufmann, 1993.

DE93a   De Jong, K. A. and Sarma, J., Generation gaps revisited, Whitley, L.D., (Ed.), *Foundations of Genetic Algorithms 2*, Morgan Kaufmann, 1993.

DO91   Dodd, N., Macfarlane, D., and Marland, C., Optimisation of artificial neural network structure using genetic techniques implemented on multiple transputers, *Transputing '91*, Vol. 2, IOS Press, p687-700, 1991

DU90   Duan, N., Mays, L.W. and Lansey, K.E., Optimal reliability-based design of pumping and distribution systems, *J. Hydr. Engrg.*, ASCE, 116(2), 249-268, 1990.

ED75   Edwards, S.F. and Anderson, P. W., *J. Phys. F*, 5, 965, 1975.

EE91   *Energy Efficiency in Offices: A Technical Guide for Owners and Single Tenants*, Energy Consumption Guide 19. UK Energy Efficiency Office, 1991.

EE94   *Introduction to Energy Efficiency in Schools*, Department of the Environment Energy Efficiency Office, 1994.

EI94   Eiger, G., Shamir, U. and Ben-Tal, A., Optimal design of water distribution networks, *Water Resour. Res.*, 30(9), p2637-2646, 1994.

EL85   El-Bahrawy, A. and Smith, A.A., Application of MINOS to water collection and distribution networks, *Civ. Engrg. Sys.*, 2(1), p38-49, 1985.

ES89   Eshelman, L.J., Caruana, R.A. and Schaffer, J.D., Biases in the crossover landscape, in [SC89, p10-19], 1989.

ES91   Eshelman, L.J., The CHC adaptive search algorithm: how to have safe search when engaging in non-traditional genetic recombination, Rawlins, G., (Ed.), *Foundations of Genetic Algorithms*, Morgan Kaufmann, 1991.

ES91a   Eshelman, L.J. and Schaffer, J.D., Preventing premature convergence in genetic algorithms by preventing incest, in Belew, R.K. and Booker, L.B., (Eds), *Proceedings of the 4th International Conference on Genetic Algorithms*, Morgan Kaufmann, 1991.

ES93   Eshelman, L.J. and Schaffer, J.D., Real-coded algorithms and interval-schemata, Whitley, D. (Ed.), *Foundations of Genetic Algorithms 2*, Morgan Kaufmann, 1993.

ES94   Esbensen, H. and Mazumder, P., SAGA: a unification of the genetic algorithm with simulated annealing and its application to macro-cell placement, *Proceedings of the 7th Int. Conf. on VLSI design*, p211-214, 1994.

FA94   Fausett, L., *Fundamentals of neural networks*, Prentice-Hall International, 1994.

FE88   Feder, J., *Fractals*, Plenum Press, New York, 1988.

FI84   Fitzpatrick J.M., Grefenstette J.J. and Van Gucht D. Image registration by genetic search, *Proceedings of IEEE Southeast Conference* p460-464, 1984.

FO66   Fogel, L.J., Owens, A.J. and Walsh, M.J., *Artificial intelligence through simulated evolution*, Wiley, New York, 1966.

FO89   Fogarty, T.C., Varying the probability of mutation in the genetic algorithm, in Schaffer, J.D., (Ed.), *Proceedings of the 3rd International Conference on Genetic Algorithms*, Morgan Kaufmann, p104-109, 1989.

FO93   Forrest, S. and Mitchell, M., relative building block fitness and the building block hypothesis, In Whitley, L. D., (Ed.), *Foundations of Genetic Algorithms 2*, Morgan Kaufmann, 1993.

FON93   Fonseca, C.M. and Fleming, P.J., Genetic algorithms for multiobjective optimisation: formulation, discussion and generalisation, in *Genetic Algorithms: Proceedings of the 5th International Conference*, Forrest, S., (Ed.), p416-423,

Morgan Kaufmann, 1993.

FU90   Futuyma, D.J., *Evolutionsbiologie*, Birkhäuser Verlag, Basel, 1990.

FU93   Furuta, H. et. al., Application of the genetic algorithm to aesthetic design of dam structures, *Proceedings Neural Networks and Combinatorial Optimization in Civil and Structural Engineering Conference*, Edinburgh (1993), published by Civil Comp Ltd, p101-109, 1993.

FUJ90   Fujiwara, O. and Khang, D.B., A two-phase decomposition method for optimal design of looped water distribution networks, *Water Resour. Res.*, **26**(4), p539-549, 1990.

GE85   Gessler, J., Pipe network optimization by enumeration, *Proc. Spec. Conf. on Comp. Applications/Water Resour.*, ASCE, New York, p572-581, 1985.

GO87   Goldberg, D.E., and Richardson, J., Genetic algorithms with sharing for multimodal function optimisation, Genetic algorithms and their applications: *Proceedings of the 2nd International Conference on Genetic Algorithms*, p41-49, 1987.

GO87a   Goldberg, D.E. and Kuo, C.H., Genetic algorithms in pipeline optimization, *J. Comp. in Civ. Engrg.*, **1**(2), p128-141, 1987.

GO89   Goldberg, D.A., *Genetic Algorithms in Search, Optimisation and Machine Learning*, Addison-Wesley, 1989.

GO89a   Goldberg, D.E., Korb, B., and Deb, K., Messy genetic algorithms: Motivation, analysis, and first results. *Complex Systems*, **3**(4), p493-530, 1989.

GO89b   Goldberg, D.E., Sizing populations for serial and parallel genetic algorithms, in Schaffer, J.D., (Ed.), *Proceedings of the 3rd International Conference on Genetic Algorithms*, Morgan Kaufmann, p70-79, 1989.

GO91   Goldberg, D.E. and Deb, K., A comparative analysis of selection schemes used in genetic algorithms, Rawlins, G., (Ed.), *Foundations of Genetic Algorithms*, Morgan Kaufmann, p69-93, 1991.

GO91a   Goldberg, D.E. and Deb, K. and Bradley, K., Don't worry, be messy, in Belew,

R.K. and Booker, L.B., (Eds), *Proceedings of the 4th International Conference on Genetic Algorithms*, Morgan Kaufmann, p25-30, 1991.

GO93   Goldberg, D.E., Deb, K., Kargupta, H. and Harik, G., Rapid, accurate optimization of difficult problems using fast messy genetic algorithms, in *Genetic Algorithms: Proceedings of the 5th International Conference*, Forrest, S., (Ed.), p56-64, Morgan Kaufmann, 1993.

GOT89   Gottschalk, W., *Allgemeine Genetik*, Georg Thieme Verlag, Stuttgart, 3rd edition, 1989.

GOU86   Goulter, I.C., Lussier, B.M. and Morgan, D.R., Implications of head loss path choice in the optimisation of water distribution networks, *Water Resour. Res.*, **22**(5), p819-822, 1986.

GR85   Grefenstette, J.J., Gopal, R., Rosmaita, B.J., and Van Gucht, D., Genetic algorithms for the travelling salesman problem, *Proceedings of an International Conference on Genetic Algorithms and Their Applications*, p160-168, 1985.

GR86   Grefenstette, J.J., Optimization of control parameters for genetic algorithms, *IEEE Transactions on Systems, Man and Cybernetics* **16**(1), p122-128, 1986.

GR89   Grefenstette, J.J. and Baker, J.E., How genetic algorithms work: a critical look at implicit parallelism, in Schaffer, J.D., (Ed.), *Proceedings of the 3rd International Conference on Genetic Algorithms*, Morgan Kaufmann, p20-27, 1989.

GR91   Grefenstette, J.J., Conditions for Implicit Parallelism, Rawlins, G., (Ed.), *Foundations of Genetic Algorithms*, Morgan Kaufmann, 1991.

GR93a   Grefenstette, J.J., Deception considered harmful, Rawlins, G., (Ed.), Whitley, D. (Ed.), *Foundations of Genetic Algorithms 2*, Morgan Kaufmann, 1993.

GR97   Greenwood, G.W., Hu, X. and D'Ambrosio, J.G., Fitness functions for multiple objective optimization problems: combining preferences with Pareto rankings, Belew, R.K. and Vose, M., (Eds), *Foundations of Genetic Algorithms 4*, Morgan Kaufmann, 1997.

HA88   Haves, P. and Littlefair, P.J., Daylighting in dynamic thermal modelling programs: case study, *Building Services Research and Technology*, 9(4) 183-188, 1988.

HA97   Halhal, D., Walters, G.A., Ouazar, D. and Savic, D.A., Water network rehabilitation with structured messy genetic algorithm, *J. of Water Resources Planning and Management*, ASCE, **123**(3), p137-146, 1997.

HE94   Herdy, M. and Patone, G., *Evolution Strategy in Action*, Presented at Int. Conference on Evolutionary Computation, PPSN III, Jerusalem, 1994.

HI94   Hill, D.L.G., Studholme, C. and Hawkes, D.J., Voxel similarity measures for automated image registration. *Proceedings Visualisation in Biomedical Computing*, Bellingham, W.A., SPIE Press, p205-216, 1994.

HI95   Hinterding, R., Gielewski, H. and Peachey, T.C., The nature of mutation in genetic algorithms, in Eshelman, L.J., *Proceedings of the 6th International Conference on Genetic Algorithms*, p65-72, 1995.

HI96   Hill D.L.G., Studholme C. and Hawkes D.J., Voxel similarity measures for automated image registration. Automated 3-D registration of MR and CT images of the head, *Medical Image Analysis*, **1**, p163-175, 1996.

HO71   Hollstien, R.B., *Artificial genetic adaptation in computer control systems*, Doctoral dissertation, University of Michigan, Dissertation Abstracts International, **32**(3), 1510B, (University Microfilms No. 71-23, 773), 1971.

HO75   Holland J.H., 1975, *Adaptation in Natural and Artificial Systems*, University of Michigan Press, Ann Arbor, 1975.

HU79   Hunt, D.R.G., The use of artificial lighting in relation to daylight levels and occupancy, *Building and Environment*, **14** p21-33, 1979.

HU91   Huang, R. and Fogarty, T.C., Adaptive classification and control-rule optimisation via a learning algorithm for controlling a dynamic system, *Proc 30th Conf. on Decision and Control*, p867-868, 1991.

JA91   Janikow, C. and Michalewicz, Z., An experimental comparison of binary and floating point representations in genetic algorithms, in Belew, R.K. and Booker, L.B., (Eds), *Proceedings of the 4th International Conference on Genetic Algorithms*, Morgan Kaufmann, p31-36, 1991.

JO95   Jones, T., Crossover, macromutation and population-based search, in Eshelman, L.J., *Proceedings of the 6th International Conference on Genetic Algorithms*, p73-80, 1995.

KA60   Kalman, R.E. *Trans ASME J Basic Eng.* **82**(D) p35, 1960.

KA97   Kawaguchi, T., Baba, T., Nagata, R., 3-D object recognition using a genetic algorithm-based search scheme, *IEICE transactions on information and systems*, **E80D**(11), p1064-1073, 1997.

KE89   Kessler, A. and Shamir, U., Analysis of the linear programming gradient method for optimal design of water supply networks, *Water Resour. Res.*, **25**(7), p1469-1480, 1989.

KI87   Kinzel, W., Spin glasses and memory, *Physica Scripta* **35**, p398-401, 1987.

KI90   Kitano, H., Designing neural networks using genetic algorithms with graph generation system, *Complex Systems* **4**, p461-476, 1990.

KI94   Kitano, H., Neurogenetic learning: an integrated method of designing and training neural networks using genetic algorithms, *Physica* **D** 75, p225-238, 1994.

KI94a   Kim, H.J. and Mays, L.W. Optimal rehabilitation model for water distribution systems, *Journal of Water Resources Planning and Management*, ASCE, **120**(5), 674-692, 1994.

KO91   Koza, J.R., Evolving a computer program to generate random numbers using the genetic programming paradigm, in Belew, R.K. and Booker, L.B., (Eds), *Proceedings of the 4th International Conference on Genetic Algorithms*, Morgan Kaufmann, p37-44, 1991.

KO92   Koza, J.R., *Genetic Programming: on the Programming of Computers by Means of Natural Selection*, MIT Press, 1992.

KO94   Koza, J.R., *Genetic Programming II: Automatic Discovery of Reusable Programs*, MIT Press, 1994.

KO95   Kobayashi, S., Ono, I. and Yamamura, M., An efficient genetic algorithm for job shop scheduling problems, in Eshelman, L.J., *Proceedings of the 6th International*

Conference on Genetic Algorithms, p506-511, 1995.

KU93   Kuo, T. and Hwang, S., A genetic algorithm with disruption selection, in *Genetic Algorithms: Proceedings of the 5th International Conference*, Forrest, S., (Ed.), p65-69, Morgan Kaufmann, 1993.

MA83   Mandelbrot, B.B., *The fractal geometry of nature*, Freeman, New York, 1983.

MA89   Manderick, B. and Spiessens, P., Fine-grained parallel genetic algorithms, in Schaffer, J.D., (Ed.), *Proceedings of the 3rd International Conference on Genetic Algorithms*, Morgan Kaufmann, p428-433, 1989.

MA93   Maruyama, T., Hirose, T. and Konagaya, A., A fine-grained parallel genetic algorithm for distributed parallel systems, in Forrest, S., *Proceedings of the 5th International Conference on Genetic Algorithms*, p184-190, 1993.

MA95   Mahfoud, S.W., A comparison of parallel and sequential niching methods, in Eshelman, L.J., *Proceedings of the 6th International Conference on Genetic Algorithms*, p136-143, 1995.

MA96   Mahfoud, S.W. and Mani, G., Financial forecasting using genetic algorithms, *Applied Artificial Intelligence*, **10**, p543-565, 1996.

ME92   Meyer, T.P., *Long-Range Predictability of High-Dimensional Chaotic Dynamics*, PhD thesis, University of Illinois at Urbana-Champaign, 1992.

ME92a   Meyer, T.P. and Packard, N.H., Local forecasting of high-dimensional chaotic dynamics, in Casdagli, M. and Eubank, S., (Eds.), *Nonlinear Modeling and Forecasting*, Addison-Wesley, 1992.

MI91   Michalewicz, Z. and Janikow, C, Handling constraints in genetic algorithms, in Belew, R.K. and Booker, L.B., (Eds), *Proceedings of the 4th International Conference on Genetic Algorithms*, Morgan Kaufmann, p151-157, 1991.

MI92   Mitchell, M., Forrest, S. and Holland, J.H., The royal road for genetic algorithms: fitness landscapes and GA performance, in Varela, F.J. and Bourgine, P., (Eds.), *Toward a Practice of Autonomous Systems: Proceedings of the First European Conference on Artificial Life*, MIT Press, 1992.

MI93   Mitchell, M. Hraber, P.T. and Crutchfield, J.P., *Revisiting the edge of chaos: Evolving cellular automata to perform computations 7*, p89-130, 1993.

MI94   Michalewicz, Z., *Genetic Algorithms + Data Structures = Evolution Programs*, 2nd edition, Springer-Verlag, Heidelberg, 1994.

MI94a   Mitchell, M., Crutchfield, J.P. and Hraber, P.T., Evolving cellular automata to perform computations: mechanisms and impediments, *Physica* **D**(75), p361-391, 1994.

MI94b   Mitchell, M., Holland, J.H., and Forrest, S., When will a genetic algorithm outperform hill climbing? Cowan, J.D., Tesauro, G. and Alspector, (Eds.), *Advances in Neural Information Processing Systems 6*, Morgan Kaufmann, 1994.

MI95   Mitchell, M., Genetic Algorithms: An Overview, *Complexity* **1**(1), p31-39, 1995.

MI96   Mitchell, M., *An Introduction to Genetic Algorithms*, MIT Press, Cambridge, Massachusetts, 1996.

MIC95   Michalewicz, Z., Genetic algorithms, numerical optimization and constraints, in Eshelman, L.J., *Proceedings of the 6th International Conference on Genetic Algorithms*, p506-511, 1995.

MIG95   Migowsky, S., *Optimisation of the energy consumption of a building using a genetic algorithm*, University of Exeter, thesis, 1995.

MIK97   Mikulin, D.J., Coley, D.A., and Sambles, J.R., Fitting reflectivity data from liquid crystal cells using genetic algorithms, **22**(3), p301-307, 1997.

MIK97a Mikulin, D.J., *Using genetic algorithms to fit HLGM data*, PhD thesis, University of Exeter, 1997.

MIK98   Mikulin, D.J., Coley, D.A., and Sambles, J.R., Detailing smectic SSFLC director profiles by half-leaky guided mode technique and genetic algorithm, *Liquid Crystals*, to be published, 1998.

MU92   Murphy, L.J. and Simpson, A.R., Genetic Algorithms in Pipe Network Optimisation, *Research Report N° R93*, Department of Civil and Environmental Engineering, University of Adelaide, Australia, 1992.

MU92a   Mühlenbein, H., How do genetic algorithms really work? 1. Mutation and hill-climbing, in Männer, R, and Manderick, (Eds.), *Parallel Problem Solving from Nature 2*, North-Holland, 1992.

MU93   Mühlenbein, H. and Schlierkamp-Voosen, D., Predictive models for the breeder genetic algorithm, *Evolutionary Computation*, 1(1), p25-49, 1993.

MU94   Murphy, L.J., Dandy, G.C. and Simpson, A.R. Optimum design and operation of pumped water distribution system, *Proceedings Conf. on Hydraulics in Civil Engineering*, Institution of Engineers, Brisbane, Australia, 1994.

NA91   Nakano, R. and Yamada, T., Conventional genetic algorithm for job shop problems, in Belew, R.K. and Booker, L.B., (Eds), *Proceedings of the 4th International Conference on Genetic Algorithms*, Morgan Kaufmann, p474-479, 1991.

NO91   Nordvik, J, and Renders, J., Genetic algorithms and their potential for use in process control: a case study, in Belew, R.K. and Booker, L.B., (Eds), *Proceedings of the 4th International Conference on Genetic Algorithms*, Morgan Kaufmann, p480-486, 1991.

PA88   Packard, N.H., Adaptation toward the edge of chaos, in Kelso, J.A.S., Mandell, A.J. and Shlesinger, (Eds.), *Dynamic Patterns in Complex Systems*, World Scientific, 1988.

PA90   Packard, N. H., A genetic learning algorithm for the analysis of complex data, *Complex Systems* 4(5), p543-572, 1990

PE90   Penman, J.M., Second order system identification in the thermal response of a working school: Paper I. *Building and Environment* 25(2), p105-110, 1990.

PE90a   Penman, J.M. and Coley D.A., Real time thermal modelling and the control of buildings. *Proceedings Congress International de Domotique*, Rennes 27-29th June 1990.

PE97   Pearce, R., Constraint resolution in genetic algorithms, in [ZA97], p79-98, 1997.

PL50   Plackett, R.L., *Biometrika*, 37, pp149, 1950.

218

PO93   Powell, D. and Skolnick, M.M, Using genetic algorithms in engineering design optimization with non-linear constraints. in *Genetic Algorithms: Proceedings of the 5th International Conference*, Forrest, S., (Ed.), p424-430, Morgan Kaufmann, 1993.

RA91   Rawlins, G., (Ed.), *Foundations of Genetic Algorithms*, Morgan Kaufmann, 1991.

RA96   Rauwolf, G., and Coverstone-Carroll, V., Low-thrust orbit transfers generated by a genetic algorithm, *Journal of Spacecraft and Rockets*, **33**(6), p859-862, 1996.

RE93   Reeves, C.R., Using genetic algorithms with small populations, in *Genetic Algorithms: Proceedings of the 5th International Conference*, Forrest, S., (Ed.), p92-99, Morgan Kaufmann, 1993.

RI89   Richardson, J.T., Palmer, M.R., Liepins, G. and Hilliard, M., Some guidelines for genetic algorithms with penalty functions, in Schaffer, J.D., (Ed.), *Proceedings of the 3rd International Conference on Genetic Algorithms*, Morgan Kaufmann, p191-197, 1989.

RO87   Robertson, G., Parallel implementation of genetic algorithms in a classifier system, in *Genetic Algorithms and Simulated Annealing*, p129-140, Davis, L., (Ed.), Pitman, London, 1987.

RO93   Rojas, R., *Theorie der Neuronalen*, Springer, 1993.

ROS93   Rossman, L.A., *EPANET users manual*, U.S. Envir. Protection Agency, Cincinnati, Ohio, 1993.

SA83   Saul, L., and Karder M., *Phys. Rev.* E**48**, R3221, 1983.

SA97   Savic, D.A. and Walters, W.A., Genetic algorithms for least-cost design of water distribution networks, *J. of Water Resources Planning and Management*, ASCE, **123**(2), p67-71, 1997.

SC69   Schaake, J. and Lai, D., Linear programming and dynamic programming applications to water distribution network design, *Rep. 116*, Dept. of Civ. Engrg., Massachusetts Inst. of Technol., Cambridge, Mass., 1969.

SC81   Schwefel, H., *Numerical optimisation of computer models*, Wiley, New York, 1981.

SC89   Schaffer, J.D., (ed.), Proceedings of the 11th International Joint Conference on Artificial Intelligence, Morgan Kaufmann, San Mateo, p750-755, 1989.

SC89a   Schaffer, J.D., Caruana, R.A., Eshelman, L.J. and Das, R., A study of control parameters affecting online performance of genetic algorithms for function optimisation, in [SC89, p51-60], 1989.

SC92   Schulze-Kremer, S., Genetic algorithms for protein tertiary structure prediction, in Männer, R, and Manderick, B., (Eds.), *Parallel Problem Solving from Nature 2*, North-Holland, 1992.

SE62   Seuphor, M., *Abstract Painting*, Prentice-Hall International, London, 1962.

SH68   Shamir, U. and Howard, C.D.D., Water distribution systems analysis, *J. Hydr. Div.* ASCE, **94**(1), p219-234, 1968.

SH75   Sherrington, D. and Kirkpatrick, S. *Phys. Rev. Lett.* **35**, p1792, 1975.

SH83   Sharpe, R. A., *Contemporary Aesthetics*, Harvester Press, 1983.

SM93   Smith, A.E. and Tate, D.M., Genetic optimization using a penalty function, in *Genetic Algorithms: Proceedings of the 5th International Conference*, Forrest, S., (Ed.), p499-503, Morgan Kaufmann, 1993.

SP91   Spears, W.M., and De Jong, K.A, On the virtues of parameterised uniform crossover, in Belew, R.K. and Booker, L.B., (Eds), *Proceedings of the 4th International Conference on Genetic Algorithms*, Morgan Kaufmann, 1991.

SP91a   Spiessens, P. and Manderick, B., A massively parallel genetic algorithm: implementation and first analysis, in Belew, R.K. and Booker, L.B., (Eds), *Proceedings of the 4th International Conference on Genetic Algorithms*, Morgan Kaufmann, p279-286, 1991.

SP91b   Spears, W.M., and De Jong, K.A, An analysis of multi-point crossover, Rawlins, G., (Ed.), *Foundations of Genetic Algorithms*, Morgan Kaufmann, 1991.

220

SP93   Spears, W.M., De Jong, K.A., Bäck, T., Fogel, D.B. and de Garis, H., An overview of evolutionary computation and machine learning: ECML-93 European conference on machine learning, *Lecture Notes in Artificial Intelligence*, **667**, p442-459, 1993.

SP93a   Spears, W.M., Crossover or mutation?, Whitley, L.D., (Ed.), *Foundations of genetic Algorithms 2*, Morgan Kaufmann, 1993.

SR94   Srinivas, N. and Deb, K., Multiobjective optimisation using nondominated sorting in genetic algorithms, *Evolutionary Computation*, Vol. 2, 1994.

ST89   Stein, D., Spinglaser, Spektrum, der Wissenschaft – Chaos und Fractale, *Spektrum*, p146-152, 1989.

ST94   Stevens, M., Cleary, M. and Stauffer, D., *Physica* A **208**(1), 1994.

SU94   Sutton, P., Hunter, D .L. and Jan, N., *Am. J. Phys*, **4**, p1281, 1994.

SY89   Syswerda, G., Uniform crossover in genetic algorithms, in Schaffer, J.D., (Ed.), *Proceedings of the 3rd International Conference on Genetic Algorithms*, Morgan Kaufmann, p2-9, 1989.

SY91   Syswerda, G., A study of reproduction in generational and steady-state genetic algorithms, Rawlins, G., (Ed.), *Foundations of Genetic Algorithms*, Morgan Kaufmann, 1991.

TA87   Tanse, R., Parallel genetic algorithm for a hypercube, *Proceedings of the 2nd International Conference on Genetic Algorithms*, p177-183, 1987.

TA89   Tanse, R., Distributed genetic algorithms, in Schaffer, J.D., (Ed.), *Proceedings of the 3rd International Conference on Genetic Algorithms*, Morgan Kaufmann, p434-439, 1989.

TA93   Tate, D.M. and Smith, A.E., Expected allele coverage and the role of mutation in genetic algorithms, in *Genetic Algorithms*: *Proceedings of the 5th International Conference*, Forrest, S., (Ed.), p31-37, Morgan Kaufmann, 1993.

TO77   Toulouse, G., *Commun. Phys.*, Juno 1977.

TO87   Todini, E. and Pilati, S., A gradient method for the analysis of pipe networks, *Proc. Int. Conf. on Comp. Applications for Water Supply and Distribution*, Leicester Polytechnic, Leicester, U.K., 1987.

VA77   Vannimenus, J. and Toulouse, G., Theory of the frustration effect II – Ising spin on a square lattice, *J. Phys.* **C10**, p537-542, 1977.

WA84   Walski, T.M., *Analysis of water distribution systems*, Van Nostrand Reinhold Co., Inc., New York, 1984.

WA85   Walski, T.M., State-of-the-art pipe network optimization, *Proc. Spec. Conf. on Comp. Applications/Water Resour.*, ASCE, New York, p559-568, 1985.

WA93   Walters, G.A. and Cembrowicz, R.G., Optimal design of water distribution networks, Cabrera, E. and Martinez, F., (Eds.), *Water supply systems, state of the art and future trends*, Computational Mechanics Publications, p91-117, 1993.

WA93a   Walters, G.A. and Lohbeck, T., Optimal layout of tree networks using genetic algorithms, *Engrg. Optimization*, **22**(1), p27-48, 1993.

WA96   Wanschura, T., Coley, D.A. and Migowsky, S., Ground-state energy of the $\pm J$ spin glass with dimension greater than three, *Solid State Communications*, 99(4), p247-248, 1996.

WH89   Whitley, D., the GENITOR algorithm and selection pressure: why rank-based allocation of reproductive trials is best, in Schaffer, J.D., (ed.), *Proceedings of the 11th International Joint Conference on Artificial Intelligence*, Morgan Kaufmann, San Mateo, 1989.

WH92   Whitley, L. D. and Schaffer, J. D., (Eds.), *COGANN-92: International Workshop on Combinations of Genetic Algorithms and Neural Networks*, IEEE Computer Society Press, 1992.

WH93   Whitley, L.D. (Ed.), *Foundations of Genetic Algorithms 2*, Morgan Kaufmann, 1993.

WH95   Whitley, L.D. and Vose, M., (Eds.), *Foundations of Genetic Algorithms 3*, Morgan Kaufmann, 1993.

222

WH95a   Whitley, D., Mathias, K., Rana, S. and Dzubera, Building better test functions, in Eshelman, L.J., *Proceedings of the 6th International Conference on Genetic Algorithms*, p239-246, 1995.

WO87   Woodburn, J., Lansey, K. and Mays, L.W. Model for the Optimal Rehabilitation and Replacement of Water Distribution System Components. *Proceedings Nat. Conf. Hydraulic Eng.*, ASCE, 606-611, 1987.

WO93   Wood, D.J. and Funk, J.E., Hydraulic analysis of water distribution systems, in *Water supply systems, state of the art and future trends*, E. Cabrera and F. Martinez, Eds., Computational Mechanics Publications, p41-85, 1993.

WR31   Wright, S., Evolution in Mendelian populations, *Genetics*, **16**, p97-159, 1931.

WR91   Wright, A.H., Genetic algorithms for real parameter optimization, Rawlins, G., (Ed.), *Foundations of Genetic Algorithms*, Morgan Kaufmann, p205-218, 1991.

YA84   Yates, D.F., Templeman, A.B., and Boffey, T.B., The computational complexity of the problem of determining least capital cost designs for water supply networks, *Engrg. Optimization*, 7(2), p142-155, 1984.

YA93   Yang, F. and Sambles J.R., *J. Opt. Soc. Am. B*, **10**, p858, 1993.

YA93a   Yang, F. and Sambles J.R., *Liq. Cryst.*, **13**(1), 1993.

YA95   Yamada, T. and Nakano, R., A genetic algorithm with multi-step crossover for job-shop scheduling problems, *Proceedings of First IEE/IEEE International Conference on Genetic Algorithms in Engineering Systems: Innovations and Applications, GALESIA '95*, p146-151, 1995.

YA95a   Yamamoto, K. and Inoue, O., Applications of genetic algorithms to aerodynamic shape optimisation, AIAA paper 85-1650-CP, 12th AIAA Computational Fluid Dynamics Conf., CP956, San Diego, CA, June 1995, p43-51.

YA98   Yang, G., Reinstein, L.E., Pai, S., Xu, Z., Carroll, D.L., A new genetic algorithm technique in optimization of prostate implants, accepted for publication in the *Medical Physics Journal*, 1998.

YO74   Young P., Recursive approaches to time series analysis. *J. Inst. Mathematics and its Applications*, p209-224, May/June 1974.

ZA97   Zalzala, A.M.S. and Fleming, P.J., *Genetic Algorithms in Engineering Systems*, IEE, London, 1997.

INDEX

A bold page number indicates a dedicated section.